SIMPLE
PLANT
PROPAGATION

SIMPLE
PLANT
PROPAGATION

Edited by

JANE COURTIER

WARD LOCK LIMITED · LONDON

ACKNOWLEDGEMENTS

The publishers gratefully acknowledge the following agencies for granting permission to reproduce the colour photographs: the Harry Smith Horticultural Photographic Collection (pp 11, 22, 50, 55, 78 and 86); and Pat Brindley (pp 26, 43 and 67). All the remaining photographs (pp 2, 15, 42, 90 and 91) were taken by Bob Challinor.

First published in Great Britain in 1987
by Ward Lock Limited, 8 Clifford
Street, London W1X 1RB
An Egmont Company

House editor Denis Ingram

Text filmset in Bembo by
Paul Hicks Limited
Middleton, Manchester

Printed in Portugal

British Library Cataloguing in Publication Data

Simple plant propagation.—2nd ed.
  1. Plant propagation
  I. Courtier, Jane
  635'.043   SB119

ISBN 0-7063-6507-0

Frontispiece: Heated propagator base accommodating three individual seed trays.

Contents

PREFACE

Even the casual observer cannot fail to be impressed by the amazing diversity of plant life. This variation has largely been brought about by the ceaseless struggle for existence that occurs everywhere in nature. Therefore many of the modifications or adaptions have considerable bearing on a plant's perpetuation and increase.

Familiar examples of this are the runners of the strawberry, the tubers of the potato and the rhizomes of the iris – all of which are modified stems adapted to serve as a natural means of reproduction.

Adaptations in plants often relate to seed propagation which is a common natural method. The bright colours of many flowers and the fragrance of others are designed to attract insects, so that they may carry out the vital task of pollination.

Plant cultivators throughout the ages have learned a great deal about propagation by observing nature's methods and by trying to imitate them.

J.C.

CHAPTER 1

HOW PLANTS INCREASE

There are two distinct methods of plant increase: **seed** (sexual) and **vegetative** (asexual).

Seeds often afford a cheap and convenient method of raising large numbers of plants. Such plants are usually healthy, for the risk of transmitting disease from parents to offspring is much less with seed than when vegetative methods are employed. A definite disadvantage with seed is slowness to reach maturity; seedlings may need nursing for several years before they 'pay their keep' in the garden. This applies, for example, in the case of bulbs and fruit trees. Obviously in such circumstances vegetative methods of increase should be used if possible.

Vegetative methods simply involve isolating portions of a plant – whether of the stem, root or leaves – and inducing them to grow and develop into separate individuals. The resulting plants are truly 'chips off the old block' and not distinct original specimens as are seedlings. Because the function of sex is not involved, vegetative propagation is termed asexual (Fig. 1).

Fig. 1 Natural asexual propagation in the creeping buttercup which normally increases by runners.

The new plants are part of the old and this ensures absolute uniformity and conformity of type. This is the principal advantage of the method. Another advantage is that many plants so raised develop rapidly into mature and productive plants which do not require careful nursing in the early stages that is so essential with seedlings, and can compete on better terms with weeds and plant pests.

It is helpful to note which type of propagation predominates in the various groups of garden plants. Annual and biennial flowers are, of course, normally increased by seed (Fig. 2).

For herbaceous perennials, vegetative propagation is more important but seed is also used extensively. Seed is a popular method for alpine and rock garden plants but many of these are also increased by cuttings or division. Many trees are propagated by seed though these and shrubs may be raised by both seed and vegetative methods.

Practically all our hardy fruits are multiplied vegetatively while the vegetable garden supplies an example of the almost exclusive use of seed. Greenhouse plants are increased by seed and by vegetative methods.

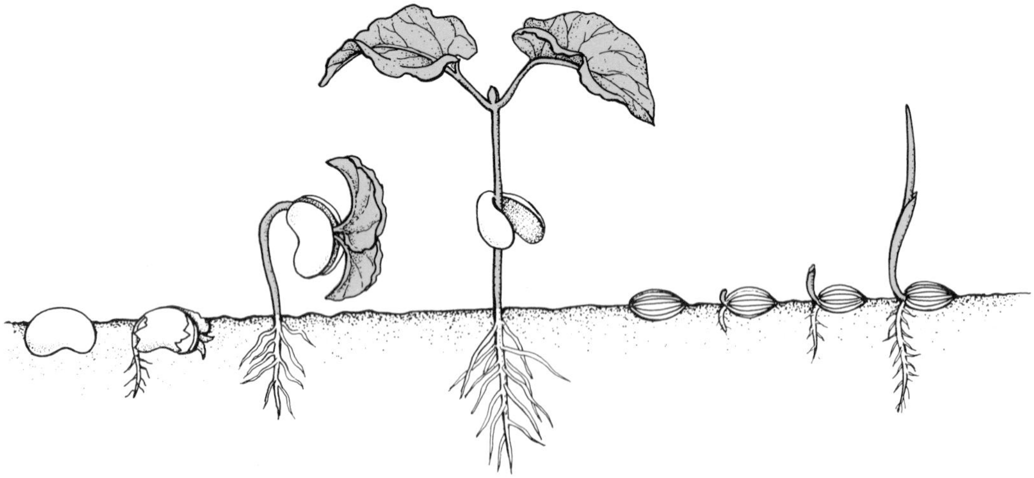

Fig. 2 The germintion of seeds. Plants may be one of two types – dicots, which produce two seed leaves, *left*, and monocots, which produce only one seed leaf.

CHAPTER 2

CONDITIONS AND EQUIPMENT NEEDED

A great many plants, whether increased from seed or cuttings, may be raised in the open. A plot situated in a warm sheltered position is most suitable for this purpose, and a sunny corner or a walled-in garden is ideal. But any plot facing south or south-west and protected from cold winds may be equally good. Ideally the soil should be light and free-working.

Greenhouses, especially if heated, greatly extend the variety and scope of propagation (Fig. 3). It is important that a greenhouse admits the maximum amount of light, particularly where most of the propagation is to be done in the short days of late winter and early spring. Good light is essential for the sturdy growth of seedlings.

An unheated greenhouse is also very useful, though it does not extend the propagating seasons as much as a heated one does.

Fig. 3 Greenhouses extend considerably the range of propagating activities you can undertake. They are essential if you want to use sophisticated mist propagating equipment.

Fig. 4 If you do not have a greenhouse,
frames are almost as useful, especially if you
make use of modern aids like coaxial soil- and air-warming cables.

Frames in the open have many uses in propagation. They can fill almost the same role as an unheated greenhouse (though on a smaller scale) but are more economical (Fig. 4).

Frames enable the propagator to provide 'close' humid conditions and are best placed in a sheltered position. Sometimes frames in the open come with a double cover consisting of a sheet of polythene over which one or more glass 'lights' are placed, according to the length of the frame. So-called 'Dutch lights' are ideal for this purpose.

Softwood cuttings from herbaceous plants and shrubs can be raised in frames during the warmer months. In the autumn semi-mature or hardwood cuttings such as conifers may be planted in cold frames. Frames are also useful for starting early vegetable crops and other plants from seed.

Propagators can be incorporated on a greenhouse bench or even in a well-lit room. Different models are available, heated by electricity or paraffin. Simple propagating units can be made, using soil-warming cables laid in sand to provide 'bottom heat'. Make sure that properly insulated cable and electrical fittings are used.

Mist units are expensive, but very useful for rooting large numbers of softwood cuttings. Nozzles are set up among the cuttings together with a sensor: when the sensor dries out a fine overhead mist of water is automatically delivered from the nozzles.

Selection of containers including plastic, clay and whale-hide pots and plastic seed trays.

Fig. 5 A step-by-step sequence to illustrate seed sowing: (a) Fill tray with compost, (b) Firm compost with presser, (c) Water the compost, (d) Shake out the seeds, (e) Sieve to evenly cover seeds, (f) Cover with glass and newspaper.

Seed trays are the most useful items on the propagator's list. Standard wood or plastic trays are 35 cm × 23 cm (14 in × 9 in) and 6–8 cm (2½–3 in) deep, and may be used for seed raising or rooting cuttings (Fig. 5). Boxes in other dimensions are available and may prove to make better use of space in some cases.

Pots come in a wide range of sizes and plastic is preferable to clay. Half-pots and pans are shallow versions and are also available in many sizes. Boxes are used to raise moderate or large numbers of seedlings or cuttings, such as annual flowers and chrysanthemums, while pots and pans are ideal for small quantities of either cuttings or seed.

Peat and fibre pots, either single or in strips, extend the range of containers for seeds or cuttings. These are planted complete, and subsequently decompose, reducing the transplanting 'check' to young roots.

Equipment used for various operations includes most of the ordinary garden implements, such as spade, fork, rake and hoe. A trowel is very necessary for planting, and one of the most important tools is the dibber. Pencil-thickness is a useful size for greenhouse work while larger dibbers for use outdoors can be about 1 inch in diameter. A good knife is traditionally the propagator's best friend: strongly made, straight-bladed, sharp, and with a handle that provides a good grip. A special knife is used for budding. Secateurs (pruning shears) are also needed. For sowing, wooden pressers, made to fit trays or pots, level and firm the compost in containers. A watering can with a fine rose is essential and a small hand sprayer is useful. Last, but of equal importance, comes a good supply of labels – plastic or aluminium are best – and a waterproof pencil.

CHAPTER 3

SAVING SEED

Can the gardener save his own seeds and get good results from them? This is possible but there are complications, and with some kinds of plant it is definitely inadvisable. Many garden plants are varieties which have been selected and propagated vegetatively to preserve a naturally occuring variation – variegated foliage, for example, or a different flower colour. Their seed would produce 'normal' plants.

Hybrid varieties will not breed true from seed either – these have been produced by crossing two different parents under controlled conditions. Plants which are allowed to set seed naturally may have been cross pollinated, so the resulting seedlings would be different varieties again.

Finally, the weather often prevents seed from being properly ripened and dried.

In the case of vegetables, it is advisable to restrict home seed-saving to those that ripen their seed early and are relatively easy to harvest. Such crops include peas, broad beans, runner beans, French beans and tomatoes.

Many shrubs are true species and will breed true from seed. Examples are several brooms, tree lupins, berberis species (excluding B. × stenophylla), cotoneasters and Eccremocarpus scaber. Saving seed from alpine plants is both interesting and successful as many of these are also true species and are easy to raise from seed.

The innumerable hybrids of herbaceous plants cannot be reproduced true to type from seed even if self-pollinated. Excellent seedlings can, however, be raised from many of these plants in a variety of colours. Examples are dahlias, lupins and delphiniums. Plants chosen for seed collection may require careful watching as the seed is often discharged on ripening. Brooms, laburnum and tree lupins are examples. Many shrubs such as berberis, cotoneasters and the snowberry produce their seeds inside berries. When extraction is done it is usually advisable to sow at once.

Further examples of herbaceous perennials from which plentiful seed can be collected – though results may be variable – are sidalcea, platycodon, campanulas, hemerocallis, alstroemeria, anthericum, oenothera, various poppies and verbascum. Coreopsis and gaillardia are normally raised from seed and are best treated as biennials.

Seed of various annuals, such as clarkia and larkspur, may be saved, but there is always a risk of cross-pollination and deterioration of the stock. Sweet pea seed is often home-saved successfully. Quite good results may be secured with seed saved from various biennials, such as sweet William, wallflowers and Canterbury bells.

With the exception of seed from certain succulent fruits including some berried shrubs, all seeds should be thoroughly dried. This may be accomplished by spreading them on sheets of paper laid in shallow containers. A warm greenhouse is an excellent place for drying seed, but sunny windows or airy sheds will also serve. When dried, each lot should be carefully cleaned. A fairly fine sieve is useful for this purpose, and this, together with the simple method of gently blowing on the seed, gets rids of most waste material. Be sure to label each batch when it is collected and write this on the packet or envelope in which the seed will go when dry and clean. Packets should be left unsealed to allow free aeration. The manner of storing seed greatly affects its length of life – ideal is a cool, dry, gently ventilated place.

Always sow seed in pots thinly: sowing lettuce seed by hand into pot of compost.

CHAPTER 4

SEED GERMINATION

Germination is brought about by allowing seed to have moisture, air and warmth. Perhaps the first is the most obvious requirement since dryness is associated with dormancy, and absorption of water by the seed is a necessary preliminary to germination. Some seeds, such as peas, beans, beetroot may be soaked in water for a period before sowing. This enables them to absorb water more rapidly than they could in the soil, so germination is accelerated. Most seed, however, requires only a limited steady supply of moisture, such as is present in moist soil, and is likely to be injured by being steeped in water.

The importance of oxygen for germinating seed is not always fully appreciated. From a practical point of view, the gardener must avoid burying seeds so deeply in the soil that air is not freely available to them. Entry of air may also be restricted in wet sticky soil, while heavy overhead watering may cause the soil to run together or cake on the surface, practically excluding air. Under such conditions the seed may be killed or give rise to poor weak seedlings.

Seed sown outside in early spring germinates slowly owing to the coldness of the soil and air. Some seeds, marrow for example, will very likely not germinate at all for they require a minimum temperature of about 13°C (55°F). On the other hand, mustard seed will sprout slowly even when the temperature is near freezing point. This indicates the wide variation that exists. Up to a limit increasing the temperature accelerates germination; its effect is very obvious when seed is sown in a warm greenhouse. Each kind of seed, however, has an optimum temperature at which it will germinate most quickly and reliably. This temperature is generally about 6° C (10° F) higher than the optimum for normal growth of the developed plant.

Seed should not be sown outside until the soil is warm enough to allow germination within a reasonable time. If, for instance, beet seed is sown in February it may decay before germination can occur. Other seeds (broad beans, parsnips and lettuce) may germinate successfully at that time. In the greenhouse, the temperature can be regulated to suit the particular kind of seed.

The germination of most seeds is not affected by light, but light is vital to all seedlings immediately their leaves are spread out above the soil. If it is absent or not of sufficient intensity, seedlings remain pale and rapidly become weak and drawn up. For strong growth good light is essential: it should not be one sided or seedlings will bend towards it. If a one-sided light source is unavoidable, seedlings should be turned daily.

The correct time of year for sowing varies from plant to plant, but the majority are best sown in spring. Some seeds, however, should be sown in autumn. They may germinate straight away and overwinter as young plants, or the seed may need periods of cold to break its natural dormancy and enable spring germination to take place. This cold treatment is called stratification and is necessary for many trees and shrubs.

While most seeds germinate within a matter of weeks, some may take a very long time – up to two years. Always check the likely germination time before discarding any 'failures'.

CHAPTER 5

SEED SOWING IN THE OPEN

Most vegetables and a good number of ornamental plants can be raised from sowing seed in the open ground. The most important point is to get the soil in the right condition for sowing.

SOIL PREPARATION

Soil should be well dug in the autumn, removing perennial weed roots, and left rough for the frost to break down the surface. The majority of outdoor sowing is done in spring, when the weather is warming up and as soon as the soil can be worked.

When the soil surface is dry and can be walked on without it sticking to your boots, it can be prepared for sowing.

A good rake is essential for producing a seed bed. It should be fairly wide, and light to handle. Once weeds have been removed, begin raking to bring the soil level and to the right texture. Remove stones and break up clods by hitting them with the back of the rake. Work from the side of the bed and rake back and forth with light strokes until the surface is level and consists of small even crumbs of soil, with no large clods or stones.

Some seeds are sown direct where they are to grow and mature; others are sown in a seed bed and the seedlings are transplanted later. They should all be sown in rows (drills) to enable you to distinguish between crop and weed seedlings.

Drills can be drawn with the edge of a draw hoe against a straight, taut line, or for short rows the handle of the rake can be pressed into the soil. Drills for most seeds should be 1–2 cm (½–1 in) deep.

Sow the seeds thinly and evenly in the base of the drill, either by tapping them carefully out of the packet or by taking pinches of seeds between finger and thumb and sprinkling them along the drill. Alternatively, small pinches of seeds can be sown at intervals along the drill, these intervals being the distance at which the plant should eventually be spaced; this method makes expensive seed go further.

Seeds of tender plants will benefit from the use of polythene tunnels as shown in Fig. 6.

Large seeds, like runner beans, can easily be sown individually at the right spacing, but remember to allow for some failures. Sow some extra seeds at the end of the row to provide plants to fill in any gaps. Sometimes small seeds can be obtained in pelleted form, where each seed is coated with an inert, clay-like substance to make it larger and easier to handle and space. Pelleted seed is expensive and not always successful in all conditions.

Once the drill is sown, pull the soil back over the seed with a rake, covering the drill evenly. Walk up the line of the drill to firm it, then label it clearly.

If the soil is very dry at the time of sowing, water the base of the drill before sowing the seeds; watering afterwards can wash the soil off the seeds. If a dry spell continues for more than a few days, however, water the drills using a very fine sprinkler. In the unlikely event of it failing to rain for some time, the seedlings will need watering, too, once they show through the soil. Water thoroughly so that it penetrates at least 2 cm (1 in) into the soil, and does not just dampen the surface.

Thin seedlings to their correct spacing when they are large enough to handle easily, but before they become overcrowded. Select the strongest seedling at each position and carefully pull away all the others around it. Thinning can be done in two stages: if the final spacing of plants should be 15 cm (6 in), say, thin them at first to 8 cm (3 in) and give them their final thinning after another couple of weeks.

TRANSPLANTING

Plants such as brassicas, leeks and biennial flowers like wallflowers are sown in a seed bed, thinned and allowed to grow into sturdy young plants, and are then transplanted to their final positions at a convenient time.

Fig. 6 Polythene tunnels are easily constructed and form an ideal way of getting vegetable seeds off to an early start.

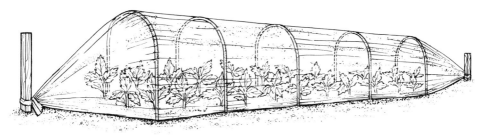

Transplanting is best done in cool, showery weather to prevent wilting. Prepare the positions for the plants first. Water the seed bed throughly, then lift the plants carefully by prising them up with a garden fork. Avoid damaging the roots as much as possible. Discard any small, weak or malformed plants.

Set the young plants out immediately in their new positions at the correct spacing, using a trowel and firming them in well. Water them in very thoroughly after planting. In hot weather, transplanting is best done in the evening.

BROADCAST SOWING

Although most seeds are best sown in drills, on occasions seed is sown at random over an area: this is known as broadcasting. Patches of annuals can be sown like this, as can seeds in seed beds for transplanting later, though weeds are often a problem with broadcast seeds. Grass for lawns is always broadcast, however.

The soil should be prepared in the same way as for any other sowing, and the grass seed should be scattered evenly over the surface. It should be sown thickly, at 2–3 oz per square yard. For accurate sowing, mark out a square metre (yard) with canes before you begin, and scatter the right amount of seed in it to help you judge the right density.

After sowing on small areas you can sieve some fine soil over the grass seed. On larger lawns, where this is impractical, rake the seed lightly into the soil surface.

CHAPTER 6

RAISING SEED UNDER GLASS

Sowing seed in greenhouses and frames gives you the opportunity to create your own climate to suit the type of plants you are propagating. They are protected from bad weather and from various pests, and uniform temperatures are easier to maintain.

A number of vegetables can be given a head start if they are sown under glass very early in the season. They can then be planted out when the weather is warmer, resulting in earlier, higher yielding crops. Lettuce, brassicas (the cabbage family), leeks and onions are good crops for this method.

Many half-hardy plants must be raised under glass, and are planted outside when the danger of spring frosts is over and growing conditions are more favourable. Marrows, tomatoes and many half-hardy annual flowers provide examples. True greenhouse plants must, of course, be propagated in a heated greenhouse. Many seeds may be germinated in a warm room, but they must be moved to a well-lighted window immediately they are through the soil.

While plants raised under cover escape the ravages of many pests, danger lurks even in a greenhouse. Infestations of both pests and diseases can spread quickly in a greenhouse and strict hygiene is essential. All structures should be washed down annually and thoroughly sprayed with a good disinfectant. Leave no rubbish lying about to provide cover. Treat pests with an insecticide as soon as they are seen. Water can also be a means of infecting plants with disease, and great care should be taken to ensure a clean supply.

In the greenhouse, seedlings should be raised in sterile compost. Garden soil should not be used because it may contain various pest and disease organisms as well as weed seeds; it is also more difficult to handle.

Seed composts contain less plant foods than potting composts: too much fertilizer can scorch delicate seedling roots and impair growth. Loam-based (John Innes type) composts can be used but soilless or peat-based composts are more popular and usually give the best results.

They are light and relatively clean to handle, and are more convenient to store than loam-based composts. They do not contain as

Large seeds, such as sweet peas, can be sown in individual pots.

much plant food as loam-based types, so supplementary feeding may need to be started earlier.

Plastic seed trays and pots should be thoroughly washed and dried before use. Wood trays and clay pots are more difficult to keep in an ideal hygienic condition, which is why they are rarely used now.

Fill trays to the brim with compost, paying special attention to the corners, and level and firm gently with a presser. (Use the base of another pot to level the surface of a pot of compost.) Water thoroughly with a fine rose on the can and leave to drain before sowing.

Very small seed, such as begonia seed, should be sown on the compost surface and left uncovered. The seeds are so tiny that they are difficult to handle, and mixing them with a small amount of fine silver sand makes them easier to sow evenly. Stir them into the sand, and shake the seed and sand mix evenly over the compost surface in a thin layer.

Most seeds need covering after sowing. Spread the seeds thinly over the compost, either by sprinkling pinches from the finger and thumb or by tapping them carefully out of the seed packet. Thick sowing results in an overcrowded mass of weak seedlings which will be prone to various fungal diseases.

Cover by sifting a little compost over the seed, the depth depending on the size of the seed. As a general rule, cover the seed to more or less its own depth with compost or silver sand.

When sowing is completed each pot is labelled. It is usual to place all the containers together and cover them with sheets of glass and paper to prevent drying out. Examine daily for signs of germination and at the same time wipe condensed moisture from the glass. Immediately germination occurs the seedlings must be placed in full light: seedlings grown in shade soon become pale and spindly and prone to disease. They must not be put straight into bright, direct sunshine, however, as this will quickly scorch and dehydrate them: diffuse light, through some form of light shading, is the aim.

If seeds are sown on moist compost and germination occurs within a reasonable time, watering should not be necessary until the seedlings are above the soil. If the compost dries before germination, the container should be carefully dipped in water and allowed to soak up moisture. After germination, water can be given when required using a can with a fine rose or a spray, and in hot weather this will probably be necessary at least once a day.

The next stage is to transplant the seedlings, which is usually called pricking on (Fig. 7). Seed trays may be used for this purpose, filled as described earlier. As a rule potting compost, which is richer in plant foods, is used. Stronger growing plants, such as tomatoes, are usually transferred from the seed tray singly into small pots.

Transplanting seedlings is necessary to allow them space for development, but it does cause a check to growth, hence the fewer moves a seedling has the better. Experiments have shown that the younger a seedling is, the less check it suffers on pricking on, so this should be done as soon as the seedlings can be handled. Tomatoes, for instance, can be transplanted the second day after they appear above the soil.

Lift seedlings carefully in order to reduce root damage to the minimum and handle them by their seed leaves only – never by their stems. Average spacing in the tray is 4–5 cm (1½–2 in) but strong-growing kinds may need a little more space.

Space them evenly in straight lines. After pricking on, seedlings need rather warm and humid conditions and may require shading for a day or two. As a rule, hardy and half-hardy plants raised in the manner described are planted outside directly from the tray or pots after being hardened off.

With true greenhouse plants the next move is usually singly into pots or from small pots to larger ones (Fig. 8). With each move a richer compost can be used.

As already explained, a cold frame often serves as an intermediate stage between the heated greenhouse and the open garden. Use a frame also for starting early vegetables like cauliflower, onions and leeks from seed, as well as raising a number of flowers. Sowing may be done directly in the frame soil or in trays or pans placed in the frame. Unless the frame is heated, do not sow before March. Later sowing usually means better and more rapid development.

7

Fig. 7 Pricking on seedlings. Both the dibber and the lifting tool can readily be improvised.

Fig. 8 Jiffys are compressed peat pots contained in a thin polythene mesh. They expand to full size once well soaked.

8

CHAPTER 7

ANNUAL AND BIENNIAL FLOWERS

Annual plants are those which grow, flower, produce seeds and die in one season. They are divided into two groups – hardy annuals and half-hardy annuals. Hardy annuals will stand some frost, and can be sown direct in the open, where they are to flower. Half-hardy annuals are not frost hardy and are usually sown in a warm greenhouse, being planted outside once the risk of frost is over.

HARDY ANNUALS

In mild districts, these may be sown in the autumn. This should result in earlier and better flowering than with spring sowing. In cold districts, however, autumn sowing is not likely to be successful as the young seedlings may be killed during the winter, so it is safer to wait until spring.

Annuals that have a relatively good chance of survival from autumn sowing are *Centaurea cyanus* (cornflower), *Iberis* (candytuft), *Delphinium ajacis* (larkspur), *Nigella damascena* (love-in-a-mist) and *Calendula officinalis* (marigold). Among others *Clarkia elegans*, *Godetia grandiflora*, *Gypsophila elegans* and *Centaurea moschata* (sweet sultan) are best sown in spring.

The soil should be in good condition. For spring sowing it should be turned over in the autumn and left exposed to the winter weather. Ensure fertility by application of fertilizers; a dusting of superphosphate raked in before sowing is beneficial. Lighten heavy soils by the addition of leaf mould, well rotted garden compost, peat and sand. Prior to sowing, rake the surface fine and level. It is usual to have a border or bed devoted entirely to hardy annuals.

Mark out the areas for each variety and label them. Draw shallow drills within each area and sprinkle the seed thinly into these drills; rake soil over to cover them and firm lightly. This is a better method than sprinkling seed at random over the whole area as it enables you to distinguish between weed and flower seedlings – the ones growing in rows are the ones to keep!

When the plants are well through the soil they must be thinned out to 15–23 cm (6–9 in) apart. Those sown in early autumn are thinned out,

either in late autumn, or left until the early spring. Thinnings are used to fill in gaps. Weeds must be kept down in the early stages.

HALF-HARDY ANNUALS
Widely used for a summer display, and huge quantities are raised every year in nurseries for sale as bedding plants. However, it is cheaper to raise your own from seed sown between January and March in a heated greenhouse. Early sowing is necessary with such kinds as antirrhinums, but many others develop more quickly and may be sown later. Sowing is as normal in trays or boxes, pricking out the seedlings as soon as they can be handled.

Gradually, as the plants develop, more ventilation is given, and then the trays are transferred to a cold frame if one is available. Here the hardening-off process continues until finally the top of the frame (the 'light') is removed, a week or two in advance of planting out. Do not plant out half-hardy annuals, however, until the risk of frost is past. Examples of this group of flowers are lobelia, petunia and zinnia.

BIENNIALS
Biennials are plants which germinate and grow in one year; then overwinter and flower, seed and die the following year. Some garden plants, such as wallflowers, are really perennials, but are treated as biennials because they deteriorate in their third year.

Biennials or flowers treated as such comprise a very useful group for spring and early summer flowering. Seed is usually sown in the open from late spring all through the summer; early sowing gives the plants a chance to become well established before the winter. Sow in drills drawn 30 cm (12 in) apart. If the soil is dry, water the open drills before sowing.

When the seedlings are large enough to handle they are pricked off in rows or in beds 15–23 cm (6–9 in) apart. In late summer or autumn set the plants in their flowering quarters. Examples are Canterbury bells, foxglove, sweet William and wallflower.

A greenhouse enables annuals to be raised in the early spring.

CHAPTER 8

SIMPLE DIVISION

Splitting up plants into several smaller ones, each with roots attached, is known as simple division. Plants that lend themselves to this method are those with tufted or matted habit. Alpine plants provide many examples of this type, including aubrieta, arabis, many dianthus and veronicas. Division is also a popular method of increasing perennial flowering plants such as Michaelmas daisies, helianthus, peonies, rudbeckia, doronicum, pyrethrum and scabious.

Plants may be divided in autumn or spring. Spring division generally gives the best results, but autumn division is useful for spring-flowering plants such as doronicums. Several hardy herbaceous perennials, such as Michaelmas daisies, do not object to being transplanted during open weather in winter.

Where plants have made large clumps with a woody centre, the young portions around the outside should be selected and the older centre discarded. Some plants are difficult to tear apart, but this is made easier with a handfork. Quite a few can be chopped up neatly with a sharp spade or a knife. Large clumps can be pulled apart by using two garden forks back to back (Fig. 9). Replant the divisions as quickly as possible, spacing them 30–60 cm (1–2 ft) apart, depending on their ultimate size and vigour.

SUCKERS
Some shrubs produce suckers – shoots which arise directly from the roots, often at some distance from the parent plant (Fig. 10).

Some suckering shrubs may be increased by simple division, including *Kerria japonica*, several of the berberis such as *B. stenophylla*, spiraeas and symphoricarpos. These may be lifted, divided and replanted in spring or autumn. Any shoot with roots attached will soon make a new plant.

Raspberry roots spread in the soil a considerable distance from the parent cane. They produce buds at various points, and these develop into strong shoots which can be severed from the parent, lifted and replanted to form a new plant. They are usually so numerous as to be a nuisance.

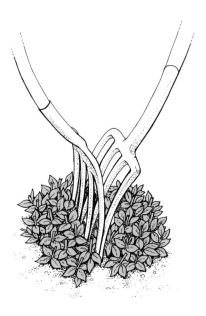

Fig. 9 Division of hardy perennials is a simple matter of placing two forks back to back and pulling the plant apart in the middle.

Fig. 10 Many woody plants can be increased by suckers. These are best detached, potted up and grown on before planting out.

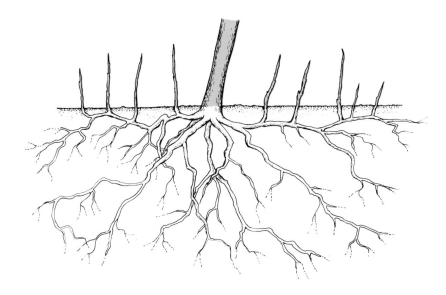

RUNNERS

These are slender stems which grow from the parent plant along the ground. The roots and shoots of a new plant arise at each node and soon become established in the soil. The runner stem eventually rots, separating the new plant from the parent. Runners can be encouraged to root rapidly by pegging them down into the soil, or by placing a stone on the runner stem near each plantlet. This ensures good contact with the soil. Strawberries are most often propagated by this method.

OFFSETS

These are similar to runners, being new young plants developing singly on a short stem arising from the parent. Echeverias and sempervivums are common examples. The young plants soon develop roots at their base when they are in contact with the soil.

RHIZOMES

These are also modified stems, though they are often mistaken for roots. Each piece of rhizome with a bud attached will grow into a new plant. The German iris is a familiar example of a rhizomatous plant, with the rhizomes forming at the soil surface. Some grasses have slender underground rhizomes which give rise to new plants wherever they are broken – the notorious weed couch grass is an example.

TUBERS

Another type of underground stem is the tuber. The potato is a common example and each of the buds or 'eyes' it possesses is capable of producing a shoot and roots. Sometimes tubers are cut in pieces and each piece having an eye may give rise to a plant. Other plants that produce tubers are Jerusalem artichokes and tuberous begonias.

A clear distinction should be made between tubers and plants with tuberous roots, such as dahlias. The latter are not true tubers and can only be used for propagation if there is a 'heel' of stem attached having one or more nodes or eyes.

The tubers of such garden plants as potatoes, Jerusalem artichokes and begonias are stored over winter and planted in the spring. The latter keep well if bedded in dry sawdust or dry peat moss and kept in a frostproof store. Begonias are usually started into growth in heat.

CHAPTER 9

BULBS AND CORMS

Bulbs and corms are also classified by botanists as modifed stems and these give rise to some of our best garden plants.

Corms are solid structures with one or more buds on the topmost side. When the corm is planted the buds grow upwards and produce foliage and flowers. In doing so the food contained in the corm is used up and the old corm shrivels. As the plants grow, however, a new corm is formed at the base of each shoot. Gladioli and crocus are good examples of corms (Fig. 11).

In some species of this type (gladioli) large numbers of new buds arise on the old corm and develop into small corms about the size of peas, called spawn. These will grow into ordinary corms but take two or three years to reach flowering size. Spawn may be induced to arise by the artificial method of making two or three cuts across the base of the corm.

Fig. 11 (*a*), a typical bulb – a daffodil. (*b*) a typical corm – a gladiolus.

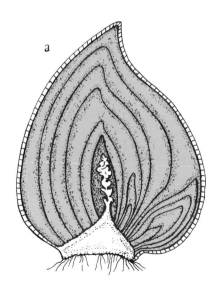
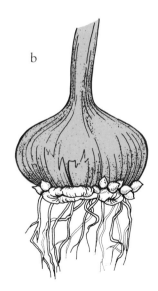

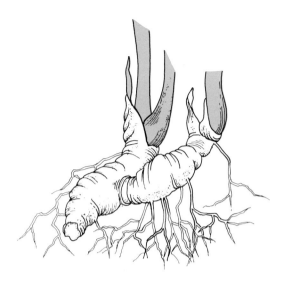

Fig. 11 (*c*) *Above*: a typical rhizome – an iris.

Fig. 12 *Right*: many lilies produce small bulbs called bulbils in the leaf axils. If potted up and grown on these will produce plants identical to the parent plant, and flower in three or four years.

A bulb has a more complicated structure than a corm and can be observed by cutting a tulip lengthwise. This is seen to consist of a short, thickened stem bearing roots on the underside and thick fleshy leaves on the upper side which encircle each other. Right in the centre and enclosed by the leaves is a large bud consisting of the undeveloped flower and foliage. Smaller buds may also be found on the short stem between the fleshy leaves.

Under suitable conditions, the central bud grows upwards and produces flowers and foliage. Small buds, if present, also develop, but few of them produce flowers. All the buds give rise to new bulbs. A large bulb usually produces several daughter bulbs but a small bulb may produce one only, which however is normally larger than its parent. This sequence of events is repeated year after year so that the number of bulbs steadily increases. Plants with scaly bulbs, such as lilies, are readily propagated by simply breaking off the scales and inserting them in a sandy compost.

Certain bulbous plants produce small or secondary bulbs called bulbils. On the flowering stalks of tree onions for instance, these are found instead of flowers. Each is capable of developing into an ordinary plant. Bulbils are also found on the stems of some lilies such as the tiger lily, *Lilium tigrinum* and the golden-rayed lily of Japan, *Lilium auratum* (Fig. 12).

CHAPTER 10

CUTTINGS

Portions of plant – from root, stem or leaves – can often be separated from the parent plant and induced to form roots and shoots of their own, growing into complete new plants. These are known as cuttings.

There are many different methods of taking cuttings, at different times of the year.

The conditions required for rooting cuttings vary according to the type of cutting (Fig. 13), but all cuttings need quite careful treatment. They have been isolated from their former supply of moisture and food, and until they grow their own roots they are very vulnerable.

Rooting powders or liquids are available to speed the rooting process. These contain substances similar to those which are produced naturally by plants, and many also contain fungicides to deter rotting of the stem. Their use is not essential but can be helpful, especially on difficult-to-root subjects. Only a light dusting of the powder is needed on the base of the stems; too much hinders rooting. Rooting hormones should *not* be used on root cuttings.

Fig. 13 Types of cuttings: (*a*) a softwood cutting, (*b*) a hardwood cutting

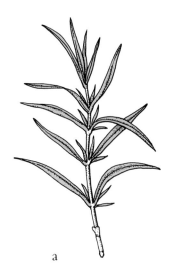

a

b

HARDWOOD CUTTINGS

These are probably the simplest cuttings to take, though you have to wait some time before you know what success you have had.

Hardwood cuttings are taken in autumn or early winter, using firm, stout growths of the current season's growth. The length of the cutting may vary from about 10 up to 30 cm (4–12 in). Side shoots can be torn away with a heel – a small piece of older wood attached to the base – otherwise the shoots can be cut just below a node (the point where the leaf joins the stem).

Two or three good buds are necessary near the top of the cutting, but buds are not required near the base. Thin, soft tips of branches should be cut off as they are likely to die back; make the cut just above a healthy bud.

Hardwood cuttings can be planted immediately or they can be bundled and half buried in moist soil in a sheltered place until early spring. They can usually be rooted quite easily in the open garden.

Select a sheltered position reasonably free from weeds. The cuttings are planted simply by putting down a line and taking out a shallow trench having a straight vertical side next to the line. Place the cuttings in the trench a few inches apart and at least three-quarters their length deep. The soil is then replaced and trodden firmly against the stems. If there is more than one row it is usual to allow at least 30 cm (12 in) between them.

The following spring, roots will grow from the bases of the cuttings and the buds will produce shoots. Leave the cuttings in place until the autumn when they can be lifted and transplanted.

SOFTWOOD CUTTINGS

These have the advantage of being quick rooting, but need a little more care. They are taken in spring and summer, using soft, leafy tips of shoots; many plants can be propagated by this method. Many herbaceous plants can be increased by basal cuttings, using the first young shoots arising from the crowns in spring, when they are 5–10 cm (2–4 in) tall.

Because the cuttings are leafy, they lose moisture rapidly from the foliage – hardwood cuttings do not have this problem. It is important to replace this lost moisture as quickly as possible, so softwood cuttings should be inserted in a moist rooting medium without delay and kept in a humid atmosphere. This usually means keeping a glass or plastic cover over the cuttings to retain moisture.

Fig. 14 The propagation of ivy from tip cuttings. A wire hoop is inserted in the pot to hold the plastic up and away from the leaves of the cutting itself.

Collect strong, healthy shoots of the current season's growth, preferably without flower buds. Cut the shoots with a sharp knife to avoid damage to the parent plant. If there is to be any delay, put the cuttings straight into a plastic bag.

Prepare a tray or pot of cuttings compost, either a commercial brand or a mixture of equal quantities of peat and sand. Trim the cuttings with a sharp knife or single-edged razor blade: usually they should be between 5 and 10 cm (2 and 4 in) long. This generally means cutting immediately below the second or third pair of leaves. Trim off these lowest leaves to give a clear length of stem; if the leaves are large, the second pair could be removed as well. If too much foliage is left on, too much water will be lost from the plant.

Dip the base of the cutting in hormone rooting powder if it is being used, and tap off the excess. Dib a shallow hole in the compost and insert the cutting. Firm it well, making sure there is good contact between the base of the cutting and the compost.

Fill a tray or pot with cuttings spaced evenly so the leaves don't touch, then water thoroughly with a fine rose on the can. Cover with a glass or plastic propagator top, or even an inflated plastic bag, and leave the cuttings in a warm, lightly shaded position in a greenhouse, garden frame, or on a window sill in the home (Fig. 14).

Rooting is usually speeded up by warmth at the base of the cutting; the amount of warmth varies according to the type of plant. Generally, the higher the optimum temperature at which a plant grows, the more warmth the cutting likes. Propagators which provide bottom heat are very useful.

Inspect softwood cuttings regularly, removing any that have died back before they start to rot. In good conditions, some cuttings can root in a matter of days.

MIST UNITS

Because of the need for a moist atmosphere, the mist unit has been developed. Spray heads are spaced at intervals over a rooting bench filled with cuttings, and a sensor (known as an electronic leaf) senses the amount of moisture in the air. When the electronic leaf dries out, the spray heads automatically switch on to deliver a fine mist of water over the cuttings. Because there is no wilting or moisture stress, softwood cuttings root very rapidly and with minimum losses in this type of unit.

SEMI-RIPE CUTTINGS

These are taken in summer, from shoots which are just begining to harden and ripen at their base. It is a popular method for shrubs: semi-ripe cuttings are less delicate than softwood but do not root quite so quickly.

Select a side shoot of the current season's growth which is firm at the base but still pliable, and tear it away from the main stem with a small heel of older wood. If the side shoot is more than 15 cm (6 in) long, cut it with secateurs, making the cutting 8–15 cm (3–6 in) in length. As with other cuttings, sever the shoot just below a node or leaf-joint.

Remove lower leaves and insert the cuttings into sandy compost, as for softwood cuttings. After watering, place the trays of cuttings in a garden frame, close it to maintain humidity, and shade the glass to prevent sun scorch. After a few days, gradually increasing ventilation can be given.

Semi-ripe cuttings may be rooted in a few weeks, but generally they need to stay in the frame over winter. They can be transplanted or potted up when growth has begun the following spring.

LEAF CUTTINGS

The leaves of certain plants may be induced to form buds and roots. Broadly speaking, propagation from leaves takes two forms. The first is exemplified by the well-known *Begonia rex* which has large ornamental leaves. In this case the leaf veins are cut through at several points. The leaf is then spread out and pegged down on damp light compost. If enclosed in a warm frame, roots and buds are produced at the incisions, each giving rise to a new plant.

A second type of leaf propagation consists in inserting the leaf stalk or a section of leaf into the compost like an ordinary cutting (Fig. 15). Strong young leaves are preferred; the stalks or leaf sections are inserted to a depth of about 4 cm (1½ in) in peaty compost. If they are kept moist and warm, new plants will arise from the compost. Plants treated in this way include streptocarpus, saintpaulia, lachenalia and sansevieria.

ROOT CUTTINGS

Buds appear naturally on the roots of such plants as raspberries and plums and give rise to suckers which are used in propagation. The roots of any such plants are also suitable for root cuttings. Moreover, there are other plants which, although they do not form such buds naturally on their roots, can be induced to do so by ordinary method of propagation. Seakale is a good example of this and most gardeners are only too familiar with the spectacle of pieces of weed roots, like dock

Fig. 15 The propagation of sansevieria by leaf cuttings. Each segment will produce a plant from its base, but the resulting plant will show no variegation until mature.

and dandelion, sprouting and giving rise to more unwanted plants.

The size of root cutting to take varies with the type of plant, but as a general rule relatively thick roots of a reasonable length are preferred (Fig. 16).

Amongst fruits, raspberries, and their relations the blackberries and loganberries, are easily increased from root cuttings. Suitable roots are 0.5–1 cm (¼ to ½ in) thick and are cut in pieces 8–10 cm (3–4 in) long. Such portions may be laid horizontally in a shallow trench and covered with a few inches of fine soil.

Various herbaceous perennials are readily increased by root cuttings. A number of these, including *Anchusa azurea* and its varieties, perennial verbascums, eryngiums and oriental poppies, may be planted in the open in the same manner as seakale. The ordinary garden phlox, *Phlox paniculata*, is often increased by root cuttings and this method is quicker than dividing the roots. It consists in selecting the stronger roots and cutting them into pieces about 5 cm (2 in) long. These are sprinkled on a sandy compost contained in boxes or pans and covered with about 1 cm (½ in) of the same medium. Place the receptacles in a frame or glasshouse and, when the shoots are well through, the young plants should be pricked off in a cold frame or in a nursery bed in the open.

Fig. 16 Root cuttings. *Left*, a diagram showing how root cuttings may be taken and, *right*, how they should be planted.

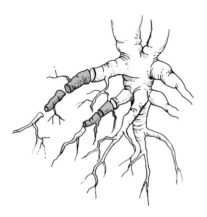 

CHAPTER 11

GRAFTING AND BUDDING

Grafting is the art of inducing a piece of the stem of one plant, called the scion, to unite with the rooted portion of another, known as the stock. The two, so united, grow together to form one plant, yet each maintains its individuality and a shoot arising from one is distinct from that produced by the other. Normally, however, the role of the stock is to provide the plant's roots while the scion furnishes the top growths.

The primary object of grafting and budding is to reproduce plants that are more difficult or cannot be propagated at all by other vegetative methods. Thus, the various tree fruits, such as apples, pears and plums, are extremely difficult to root from any type of cutting, while by grafting large numbers can be raised with comparative ease.

Another important reason for grafting is to enable certain plants to be grown on roots other than their own, and this is often advantageous. Tree fruits again provide an example and, in the case of apples, several distinct root stocks are used with the object of securing trees suitable for different purposes and conditions.

Roses are usually grafted (or budded), although most varieties root easily from hardwood cuttings.

Grafting has several disadvantages. Grafted plants may be prone to suckering, that is, to producing unwanted shoots from the root stock. This is a bad fault with some plum stocks, and can be a serious nuisance. Usually, stocks for grafting in the open are planted in the soil, and for most subjects such as fruit trees, or roses, should be established at least a year before grafting or budding is attempted.

Grafting out of doors is usually undertaken in early spring just when growth is commencing. The scions, however, should be severed from the parent plant when they are quite dormant. Usually they are then bunched together and laid in damp sand or soil in a shady position.

Success with the operation of grafting is largely a matter of bringing the cambium (an active layer just below the bark) of the stock and the cambium of the scion together, and no union can occur unless this is achieved. This, therefore, should be the aim, and it is facilitated by making firm, clean cuts with a really sharp knife in such a way that the parts fit snugly and firmly together (Fig. 17 nd 18).

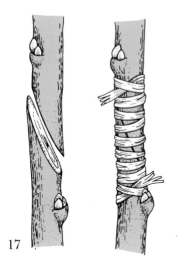

Fig. 17 *Left:* splice or whip grafting – by far the simplest and most widely used grafting technique of all.

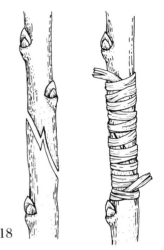

Fig. 18 *Right:* whip and tongue grafting, a variant on simple whip grafting, ensures an even tighter fit of cambium to cambium.

18

Scions vary in length according to the type of plant, but a common length is about 15 cm (6 in). The stocks should be well established and sufficiently strong to sustain the grafted plant.

With most methods of grafting, it is necessary to tie the scion firmly to the stock and raffia is often used for this purpose. After tying, the union is usually sealed around with a wax or jelly to prevent the entry of water and to check drying out of the wounded tissues.

Grafting wax may be purchased ready for use, and some products may be used cold. Petroleum jelly is also used for sealing grafts and is fairly effective. Adhesive tape can also be used both as a tie and to act as a substitute for wax.

Although budding is usually regarded as a distinct method of propagation, it is really another form of grafting. Instead of being a length of stem being used as the scion, budding consists of grafting a piece of rind with bud attached on to another stem. The union takes place between the cambium layer attached to the piece of rind and the cambium layer of the stock. Some of the disadvantages often attributed to grafting, such as suckering and the possibility of using unsuitable stocks, apply of course to budding.

Roses are the most important group of plants that are increased by budding, but it is also used in the propagation of fruit trees.

Budding is generally done at the height of the growing season, that is in early summer. This is because during these months the rind of most plants separates easily from the wood and makes the operation easier.

Stock plants should be established on the site for at least a year. Before commencing to bud, any laterals near the stock base should be trimmed off so as to have a clean stem for the bud.

In the case of roses, the bud is inserted near the ground. This tends to reduce the production of suckers and encourages the scion to produce roots also, which in this case is desirable.

Young stems of the current year's growth provide the buds. After being severed the stems should be protected from drying out. For this reason dull, moist weather is preferable for budding, and the buds should be inserted in the stock immediately they are isolated. Select plump buds from medium-sized stems and do not use those on the immature soft tips.

There are several methods of budding but by far the most important is called 'shield' or 'T budding' (Fig. 19). This is a simple operation which with a little practice anyone should be able to perform successfully. A good sharp budding knife is essential. The stock is

Fig. 19 Budding is simply a rather specialized form of grafting. The method shown here is known as 'shield' or 'T-budding'.

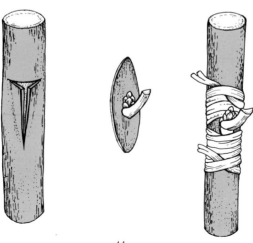

Softwood cuttings: propagation of pelargonium (*1*) Removing leaves from
lower part of stem. (*2*) Cutting stem just below a leaf node. (*3*) Inserting
cuttings at side of pot. (*4*) Mature pelargonium cuttings.

4

prepared by making a clean vertical cut about 2.5 cm (1 in) long right down to the wood, but no deeper. At the top of this, make a transverse cut to form a 'T'-shaped incision. A bud is then sliced from the stem with about 1 cm (½ in) rind on either side of it.

There is always a great deal of controversy as to whether or not the wood adhering to the rind should be pulled off. As a rule, it is better to do so because this allows improved cambial contact. In removing the wood, however, there is the risk of pulling out the base of the bud, which is a little green structure about the size of a pinhead. The attached leaf on a bud should be cut off so as to reduce transpiration to the minimum, but the leaf-stalk or petiole is allowed to remain and serve as a handle.

When the bud is ready, the rind is carefully and cleanly loosened and lifted from the wood on either side of the incision on the stock. Into this the little shield held by its tiny handle is slid down from the top. To complete the operation, bind the union firmly with raffia, leaving the bud uncovered. The binding prevents drying out and the entry of rain or insects.

CHAPTER 12

LAYERING

Layering is a method of plant propagation which is designed to induce plant stems to produce roots while they are still attached to and sustained by the 'parent' plant. It is a reliable means of increase and is often adopted for species which are difficult or impossible to root from cuttings. Layering does not lend itself to the production of large numbers of plants and is, therefore, used by the nurseryman only for the more difficult species.

This method is of particular advantage to the amateur who does not require a large quantity of any particular species. Layering is comparatively easy to perform and does not require the facilities necessary for certain other methods of propagation, such as softwood cuttings. Layering is used largely for increasing woody plants, but a few of the herbaceous type, notably border carnations, are propagated in this way.

The soil for layering around the stock plants should have liberal applications of peat and sand to make it porous and more retentive of moisture. Weeds should be kept down by cultivation as necessary. Early autumn is a good time to layer plants, but it can be done at almost any time of year.

ORDINARY LAYERING

Ordinary layering may consist of no more than bending the shoots downwards and covering the portion of the stem nearest the ground with soil.

The gardener can easily layer a few shoots of any shrub in his garden, provided the stems can be brought down to the ground. It is essential to use young shoots only. Older branches are usually slow to root and rarely develop into well-shaped specimens. A notch, slit, ring or even the paring off of some of the bark on the part to be layered is usually beneficial. Secure a good right-angle bend and fix firmly in the soil, using a peg if necessary (Fig. 20).

Most shrubs root in less than a year, and if layered during the autumn and winter are often ready for transplanting the following autumn. A layered shoot should, however, be carefully examined before severing and, if insufficiently rooted, should be left for another year.

Practically any woody plant may be increased by layering, including magnolias, pieris, and rhododendrons. Certain procumbent or horizontally growing conifers often seen in rock gardens are convenient plants to layer, and various other dwarf shrubs found on the rock garden may be layered *in situ*. These include *Daphne cneorum* and dwarf rhododendrons.

SERPENTINE LAYERING

Serpentine layering, often used on clematis, is carried out in late spring or early summer. Pliable young stems are laid along the ground and small nicks are made in the underside of the stem by each node. These are then buried, leaving the connecting portions of stem above ground in a series of loops. By autumn each buried section should have rooted and formed new plants which can be dug up and separated.

Fig. 20 Layering is one of the most certain methods of increasing most shrubs. Note that the layer is partly severed before being placed in the soil, and is then staked to prevent movement.

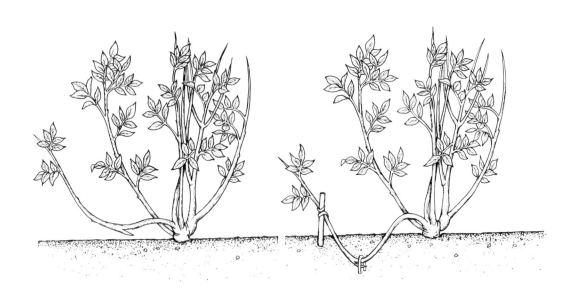

TIP LAYERING

Perhaps the simplest form of layering is tip layering which often occurs naturally and is the normal method of increasing blackberries and loganberries. Tip layering consists of bending down firm young shoots in summer and burying their tips in the soil to a depth of 10–15 cm (4–6 in). Alternatively, they may be inserted in pots. Rooting usually occurs in a matter of weeks and the new plants may be severed and transplanted in the autumn.

AIR LAYERING

In all the various forms of layering described so far, rooting is induced by bringing the stems into contact with the soil or other moist medium. There is one quite distinct type of layering, however, whereby the moist material is brought to the stem, the latter being left in its original position. This is called 'air layering' and is believed to have been used by gardeners for thousands of years. It is a useful method for propagating rare and valuable plants which are difficult or impossible from cuttings, and where ordinary methods of layering are not practicable. Thus, plants with rigid branches borne high above the ground which could not be brought down to the soil can be layered. Certain greenhouse and house plants, such as *Ficus elastica*, are propagated in this way.

Clean young shoots are best for air layering. They are prepared by making an upward sloping cut in the stem, being careful not to cut more than half way through. Alternatively, the stem may be girdled by removing a ring of bark about 1 cm (½ in) wide. The wound may be dusted with rooting powder, and a handful of moist sphagnum moss is packed between the cut surfaces and all around it, so as to give complete cover. Thin polythene is wrapped around the sphagnum moss in such a manner as to leave no opening which would allow evaporation of moisture from the moss. The film is held in position by a piece of tape tied around the stem both above and below the layer.

CHAPTER 13

PROPAGATING PERENNIAL FLOWERS

There are several ways to propagate most plants, and the methods given below are not necessarily the only ways in which the particular plants can be increased – they are just the most commonly used or most generally successful. There is no reason why other methods of propagation cannot be tried if suitable material is not available for the recommended method, but results may be variable.

HOW THE DIFFERENT KINDS ARE INCREASED

Acanthus (bear's breeches) Root cuttings are the usual method, but division in the spring is also successful. Seed is another means of increase.

Achillea (yarrow) Easily increased by division in spring or autumn. Soft cuttings strike readily in spring. The species come true from seed.

Aconitum (monkshood) Sow seed outside in mid- to late spring. Divide in spring or autumn.

Alstroemeria (Peruvian lily) Divide the roots in spring or sow seed thinly under glass in the autumn or early in spring. Seedlings must be handled with great care and disturbed as little as possible.

Anchusa (bugloss) Easily increased from root cuttings about 10 cm (4 in) long planted in the open or in a cold frame in spring: also from seed.

Anemone (wind flower) The popular. A. *hupehensis* (syn. *A. japonica*) and its varieties are usually increased from root cutting in spring. Division in the autumn is another method, but the rooted pieces should preferably be potted.

Anthemis (chamomile) Readily propagated from soft cuttings in spring or by division in spring or autumn.

Aquilegia (columbine) The species can be increased from seed if plants are kept well isolated from one another. Excellent strains in mixed colours can also be raised from seed. Sow in the open in spring. Division of named varieties may be done in spring.

Artemisia (wormwood) Mainly increased by division; also semi–ripe cuttings.

Astilbe (goat's beard) Divide during the dormant period or sow seeds under glass in late winter.

Bergenia Divide in spring or autumn.

Campanula (bell flower) The tall herbaceous types are easily increased by division or spring cuttings. Seed may also be used.

Chrysanthemum C. *maximum*, the Shasta daisy, is readily increased by division or by spring cuttings.

Named varieties of border chrysanthemums are propagated by softwood cuttings in spring. Healthy plants which are true to type and variety are selected and labelled at flowering time. After flowering, in the autumn, they are cut down to about 15 cm (6 in) above soil level and the roots are lifted and washed to remove soil. All leafy growths are also removed. The roots (known as 'stools') are packed in boxes such as tomato boxes, surrounded by peat.

During the winter the boxes of stools should be stored in a dry, cool but frost-free place.

Cutting are taken from early winter to early spring, depending on variety and the facilities available. The boxes of stools are watered to moisten the peat thoroughly, and brought into a warm, light greenhouse. Shoots are soon produced from the mass of roots, and when they are 5–8 cm (2–3 in) long they can be used as cuttings.

Snap the shoots off and trim them to about 4–5 cm (1½–2 in), cutting immediately below a node, and remove lower leaves. Dip the base of the cutting in hormone rooting powder and insert it firmly into a pot or tray of seed compost topped with a layer of sharp sand. Water well after insertion, then keep the cuttings in a warm humid position, around 15° C (60° F), covering them with a propagator top or spraying frequently with a fine mist of clear water. Once rooted they can be potted up individually or allowed to develop in the trays until they are large enough to plant outside, depending on the time of year at which they are taken.

Coreopsis (tickseed) The perennial species are raised from seeds sown in the open in spring; or divide in spring or autumn.

Dahlia These can be raised from seed sown in gentle heat in late winter or early spring, but named varieties are increased by cuttings or division of the tuberous roots.

Plants are lifted as soon as frost blackens the foliage. The tuberous roots are cleaned and allowed to dry very thoroughly before being stored away in peat in the same way as chrysanthemum stools. Any damaged portions should be trimmed off and the cut areas dusted with sulphur to prevent fungal decay. Inspect roots occasionally during the winter to ensure they are not shrivelling or rotting.

Dahlias are started into growth and cuttings taken in just the same way as chrysanthemums.

Begonia leaf cuttings: young plantlets emerging from leaf.

If you want to divide the tuberous roots, this is done in the spring. Each portion must have a section of the main stem attached, complete with a growth bud or 'eye': a good root can be split into four or five new plants. The sections are planted out in the same way as complete roots.

Delphinium Propagate from basal cuttings taken with a heel in a slightly heated glasshouse and cold frame. Seed sown in heat in mid-winter and planted out in the spring will produce plants flowering in the autumn. Seed may also be sown in spring or early summer in the open.

Dianthus (pinks and carnations) Propagate by layering in summer or taking 'pipings'. These are sections of stem pulled out from between the pairs of leaves and treated as softwood cuttings.

Dicentra (bleeding heart) Divide in spring or take cuttings at the same time.

Digitalis (foxglove) The perennial species are increased by seed or division.

Doronicum (leopard's bane) Easily increased by division in autumn.

Echinops (globe thistle) Root cuttings planted in a cold frame in spring are effective; or divide in spring or autumn.

Eremurus Propagate by division in spring. Seed is also feasible but is often slow to germinate and seedlings take three years to reach flowering age.

Erigeron (fleabane) Divide in spring or autumn or secure basal cuttings in spring.

Eryngium (sea holly) Take root cuttings in spring or sow seeds in early summer.

Gaillardia (blanket flower) Name varieties do no come true from seed and should be raised by division in spring or by root cuttings.

Geranium (crane's bill) Divide the plants in spring or autumn and sow seed in spring.

Geum (avens) The hybrids should be increased by spring division but several good strains can be raised from seed sown in spring.

Gypsophila (chalk plant) G. *paniculata* is easily raised from seed sown in heat in late winter or out of doors in spring. Varieties must be raised either from soft cuttings secured in spring from plants grown in heat or by grafting on seedlings. The varieties Bristol Fairy and *flore plena* (double form) are increased in this manner.

Helenium (sneezeweed) Can be divided almost any time during the dormant period and is also easily increased from spring cuttings.

Helianthus (sunflower) Increase is similar to *Helenium*.

Heliopsis Easily increased by division.

Helleborus (Christmas rose) Divide in spring or autumn (resents disturbance). Sow seed immediately it is ripe in a cold frame, or in the open.

Hemerocallis (day lily) Divide clumps in spring or summer.

Heuchera (alum root) Divide in spring or increase by seeds sown in spring. Seedlings, however, are often variable.

Incarvillea Increased by division or seed, which takes three years to reach flowering size.

Iris (flag) Seed is one means of increasing this large genus, and most species can be so raised. Usually seed is sown when ripe and, after exposure to winter cold, germinates readily in spring or mild heat. Division is also widely used and is the only means of increasing varieties. The bearded irises should be divided immediately after flowering.

Kniphofia The red-hot pokers are normally increased by division in spring. The species come true from seed sown in spring.

Liatris Easily increased from offsets secured and transplanted in spring. Seed may also be used, sown in spring.

Lupinus Lupins are easily increased from seed sown in heat in early spring or in the open later in the season. Named varieties are best increased from heeled cuttings secured when available in spring and potted. Lupins are difficult to divide, though it may be attempted.

Lychnis (campion) Easily increased by division in spring or autumn. Seed sown outside in spring provides another method.

Lysimachia (loosestrife) Propagate by division in spring or autumn.

Lythrum (purple loosestrife) Propagation is by division, seed, or basal cuttings in spring.

Malva (mallow) Easily increased by seed or cuttings.

Monarda (bee balm) Divide in spring or use soft basal cuttings.

Montbretia (tritonia) Easily increased by division.

Nepeta (catmint) Readily increased by division or by soft cuttings taken in summer and autumn.

Oenothera (evening primrose) Insert cuttings of the perennial species in a frame before flowering. Divide in spring. Seed is another method.

Paeonia (peony) The species should be raised from seed sown when ripe in cold frame. Sometimes germination is slow. Divide in autumn, ensuring that each piece of root has an 'eye'. Peonies resent disturbance and division retards flowering.

Papaver (poppy) The well-known oriental poppies are easily increased from root cuttings about 10 cm (4 in) long secured and planted in the open in spring.

Phlox *P. paniculata* can be raised from seed, but its many varieties are increased by basal cuttings or by division in spring or autumn.

Physalis (Chinese lantern) Easily increased by division or root cuttings.

Physostegia (false dragon head) Sow seed in cold frame in spring or divide roots in autumn or spring.

Platycodon (Chinese bellflower) Propagation is by seed or basal cuttings in spring.

Polemonium (Jacob's ladder) Divide in spring or autumn. Seed may be sown in spring.

Polygonatum (Solomon's seal) Propagate by division in spring or autumn.

Potentilla (cinquefoil) Increase by seed sown in the open in spring or divide in spring or autumn.

Pyrethrum The pyrethrum is really a chrysanthemum. It is increased by division in spring or immediately after flowering, or by basal cuttings in spring.

Left: Saintpaulia: taking a leaf cutting.

Below: Saintpaulia: young plantlets emerging from junction of leaf cutting and surface of compost.

Rudbeckia (cone flower) Divide in spring or autumn. Seed may be sown out of doors in spring.

Salvia (sage) Increase by semi-ripe cuttings or seed sown in spring.

Scabiosa Divide in spring only. Seed is another method but varieties do not come true.

Sidalcea Easily increased by division in spring or autumn.

Solidago (golden rod) Easily increased by division in spring or autumn.

Stachys Increase by division in autumn or spring.

Stokesia Usually propagated by seed or division; also by root cuttings.

Thalictrum (meadow rue) Divide in spring just as growth commences or sow seed in the open in spring.

Tradescantia (spiderwort) Basal cuttings in spring root readily or plants may be divided in spring.

Trillium (American wood lily) Increased from seed sown when ripe.

Trollius (globe flower) Propagate by division just after flowering or sow seed when ripe.

Verbascum (mullein) Readily increased from root cuttings in late winter, or division, or by seed sown in spring.

Veronica (speedwell) Divide in spring or autumn, or take basal cuttings in spring.

CHAPTER 14

PROPAGATION OF TREES AND SHRUBS

Seed is widely used in the propagation of trees and shrubs. Autumn sowing immediately the seed is ripe is often advantageous. However, as the seed is unlikely to germinate before late spring or early summer, weeds may be a problem. With such seed, therefore, over-winter stratification may be preferable. A simple method of doing this is to mix the seed with moist soil, peat or sand or a mixture of these and place in boxes or pots which are buried outside.

In the propagation of shrubs from softwood or semi-hardwood cuttings mist can be used almost without exception.

HOW THE DIFFERENT KINDS ARE INCREASED

Abelia. In mid-summer semi-ripe sideshoots with a heel and insert in a close case.

Abutilon (Indian mallow) Softwood cuttings root readily in early summer and semi-ripe cuttings may be stuck in a cold frame in mid- to late summer . *A. vitifolium* may be raised from seed sown indoors in winter.

Acer (maple) Most of the species and certain varieties may be raised from seed sown in the open as soon as ripe or stratified and then sown. *A. negundo* varieties and *A. palmatum* may be grafted on the seedling species. Softwood cuttings will root in a greenhouse or frame from mid-spring through to early summer.

Actinidia Seed can be sown in gentle heat in a greenhouse in early to mid-spring. Semi-ripe cuttings are taken in the summer and hardwood cuttings in late autumn or early winter: these should both be rooted in a greenhouse.

Ampelopsis Softwood cuttings can be struck in a close case in mid-summer, and in the autumn hardwood cuttings will root in a cold frame.

Aralia (angelica tree) Usually increased from suckers and root cuttings in spring. Seed is another method.

Aristolochia (Dutchman's pipe) Softwood cuttings in cold frame in summer will root but slight bottom heat is an advantage. Also raised from seed and semi-ripe cuttings.

Arundinaria (bamboo) Divide old plants in spring and for speedy rooting pot the divisions and keep in a warm, humid glasshouse.

Atriplex Propagate by semi-ripe cuttings. Annual varieties are raised from seeds.

Aucuba (spotted laurel) Semi-ripe cuttings usually strike readily. Layering is a still more certain method and can be recommended to the amateur.

Berberis (barberry) Many species come true from seed including *B. darwinii*, *B. gagnepainii* and *B. thunbergii* if each is reasonably isolated from other species. Stratify the seed over the winter and sow in the open in early spring. Special varieties and hybids such as *B.* × *stenophylla* do not come true from seed and must be propagated vegetatively. Semi-ripe cuttings usually strike easily in a cold frame.

Buddleia Cuttings are a popular and easy method and will strike in the open. Use ripe wood and make the cuttings 15–20 cm (6–8 in) long. Plant in mid- or late autumn. Softwood cuttings 8–10 cm (3–4 in) long also strike readily in a close case in the summer. *B. alternifolia* is often raised from seed sown in heat in late winter.

Buxus (box) Most of the boxwoods are easy from cuttings. Take semi-ripe shoots with a heel and plant in a cold frame in early autumn.

Calluna (ling) Propagate as for *Erica*.

Camellia *C. japonica* is raised from seed sown in heat in late winter. Prior to sowing it is recommended to soak the seed in warm water for 24 hours. *C. japonica* and *C. reticulata* varieties are usually grafted on seedlings of the ordinary type. This is done in a close case in summer. Several varieties including singles may be raised from cuttings of semi-ripe wood in a close case in mid-summer. Another method is leaf bud cuttings.

Campsis Increase by seed if available. Other methods are ripe cuttings, suckers and root cuttings in moderate heat.

Caryopteris Increased by softwood cuttings in early summer, or semi-ripe cuttings later in the season.

Ceanothus Semi-ripe cuttings with a heel are taken in late summer or early autumn and rooted in a cold frame. Softwood cuttings will also strike in early summer in a close case with bottom heat. Root cuttings can be taken in late autumn or early winter.

Ceratostigma (lead wort) Semi-ripe cuttings with a heel strike readily in a close case. Division is another method.

Chaenomeles The well-known *C. lagenaria* (syn. *Cydonia japonica*), Japanese quince, and other species may be raised from seed which is collected when ripe, stored dry over winter and sown in early spring. Varieties are often grafted on to seedling stocks, but plants on their own roots are superior and may be had by layering.

Chamaecyparis (cypress) The ordinary species such as *C. lawsoniana* (Lawson's cypress) are raised from seed sown in the open in spring. Special forms and varieties are raised from cuttings consisting of sideshoots with a heel. These are inserted in a cold frame in the autumn. Grafting on seedling stocks under glass is used to increase the varieties.

Chimonanthus praecox (winter sweet) Sow seed in early spring. Layering is another method often used.

Choisya ternata (Mexican orange) Cuttings of mature wood 10–15 cm (4–6 in) long with a heel will root in a cold frame when inserted in autumn. The immature tips root readily in a close case and are taken in summer.

Cistus (rock rose) True species are raised from seed sown under glass in late winter. Cuttings of sideshoots taken with a heel root readily in a close case in summer. Wood that is almost mature may be used for cuttings in autumn and planted in a cold frame.

Clematis Seed may be used when available to increase the species. Grafting can be used for the various hybrids but serpentine layering is the commonest method. Internodal cuttings of half-ripened wood taken in late summer will strike in a close case. They are often inserted singly in pots.

Clerodendrum Usually propagated from suckers, but can also be increased by root cuttings and semi-ripe cuttings.

Clianthus (glory pea) *C. formosus* is raised from seed. *C. puniceus* is raised from seed or semi-ripe cuttings about 8 cm (3 in) long, inserted in a close case in summer.

Colutea (bladder senna) The best plants are produced from seed which is kept dry until sowing time in late winter or early spring under glass. Semi-ripe cuttings may be rooted in a close case or frame in summer.

Corokia May be raised by seed in spring or by cuttings in summer.

Coronilla (crown vetch) Easily raised from spring-sown seed or soft or semi-ripe cuttings.

Corylopsis Layering is the usual method but soft cuttings in summer in the greenhouse will also strike.

Cotoneaster Seed is successful for many species including *C. bullata*, *C. frigida*, *C. lactea* and *C. simonsii*. It should be collected when ripe, stratified during the winter and sown in the open in early spring. Cuttings are used for several species, such as *C. adpressa*, *C. horizontalis*, *C. microphylla* and *C. rotundifolia*. These should always be taken with a heel and may be either ripe wood taken in late autumn and planted in a cold frame, or half-mature sideshoots inserted in a frame in mid-summer.

Cupressus (cypress) *C. macrocarpa* and *C. sempervirens* are raised from seed sown in late winter or early spring indoors, or outside later in the spring, or from semi-ripe cuttings. Grafting is also done for varieties under glass in summer, *C. macrocarpa* being used as a stock.

Cytisus (broom) Several species are easily raised from seed, but crossing occurs readily unless they are well isolated. Sow in spring using pans placed in a cold frame. Semi-ripe cuttings are used for some species and several varieties, e.g., *C. ardoinii*, *C. praecox*, *C. purgans* and *C. purpureus* and its varieties. These may be either half-mature sideshoots with a heel inserted under bell glasses in late summer or firm sideshoots secured in the autumn and planted in a cold frame. Some types such as *C. × kewensis* and *C. × beanii* are grafted on to seedling laburnum in early spring, in the open.

Daboecia Semi-ripe cuttings are taken throughout the summer and rooted in a cold frame or greenhouse. Plants layered in spring by 'dropping' – covering the centre with soil to leave just the tips of the branches showing – will produce many new plantlets.

Daphne Seed is used for several species such as *D. laureola* (spurge laurel) and *D. mezereum* (mezereon) and its varieties. Stratify the seed during the winter and sow in early spring in a cold frame. Cuttings are successful for *D. cneorum* (garland flower) *D. collina* and *D. retusa*. They consist of semi-mature sideshoots inserted in a cold frame or bell glasses in summer. Root cuttings in heat is the usual method for *D. genkwa*. Another is grafting under glass in spring. *D. laureola* being the stock for evergreen species and *D. mezereum* for deciduous types.

Deutzia All species and varieties are readily increased by softwood cuttings 10–13 cm (4–5 in) long taken in summer and inserted in a close case or frame. Hardwood cuttings about 20 cm (8 in) long are usually taken with a heel and in mild districts can be planted outside in late autumn. In colder localities they should be inserted in a cold frame and made somewhat shorter. Seed stored dry over winter will germinate in a warm glasshouse in late winter.

Diervilla (bush honeysuckle) Take semi-ripe cuttings in the summer, rooting them in a cold frame or greenhouse. Hardwood cuttings in late autumn will also root in a frame.

Eccremocarpus *E. scaber* (Chilean glory flower) is easily raised from seed, which is abundantly produced. Sow in pans and leave outside over winter. At the end of the winter bring indoors for quick germination.

Elaeagnus Deciduous species can be raised from seed stratified about 18 months and sown in early spring in heat. Softwood cuttings of *E. multiflora* and *E. glabra* will root in a close case or frame. Semi-ripe cuttings 10–13 cm (4–5 in) long with a heel can be struck in a cold frame. Layering is approved for several difficult species such as *E. angustifolia* (oleaster), *E. macrophylla* and *E. umbellata*.

Erica (heath) The usual method is by cuttings of semi-ripe tips about 4 cm (1½ in) long. Insert these in a compost of peat and sand in small pots and place in a close case. Mature cuttings taken in the autumn can also be rooted similarly. Layering is another method. Seed is used for a few species such as *E. arborea* (tree heath) and *E. lusitanica* (Portuguese heath). It should be sown on sifted peat in late winter.

Escallonia Softwood sideshoots 5–10 cm (2–4 in) long root in a close case and mature shoots of similar types can be inserted in a cold frame in the autumn.

Eucalyptus (gum tree) Easily raised from spring-sown seed in moderate heat.

Euonymus Softwood or semi-ripe cuttings of *E. japonicus* with or without a heel strike in a close case in summer. Ripened wood with a heel, planted in mid-autumn, can be rooted in a cold frame. *E. radicans* can be rooted from softwood cuttings. Division is suitable for some varieties and several species can be raised from seed sown in spring.

Fatsia (fig-leaf palm) Usually raised from seed, but cuttings also feasible.

Forsythia *F. suspensa* (golden bell) from hardwood cuttings 15–20 cm (6–8 in) long and planted in a cold frame in mid-autumn or in a sheltered site in the open. Semi-mature tips root readily in a close case in summer. Layering is another easy method.

Fothergilla (American witch hazel) Increased by layering in autumn or spring. Cuttings of semi-ripe wood will strike in late summer in moderate heat.

Fuchsia Cuttings made from soft tips when available can be rooted in a close case at any time of the year.

Garrya Layering in the autumn is the usual method. Semi-ripe cuttings 8–10 cm (3–4 in) long with a heel will root when inserted in a cold frame in the autumn.

Gaultheria Several species including *G. shallon* are raised from seed sown in late winter. Layering and division are other suitable methods. Certain species such as *G. hispida* (snowberry) and *G. oppositifolia* can be raised from cuttings secured in late summer and planted in a frame.

Genista *G. hispanicus* (Spanish gorse) and other species are raised from semi–ripe cuttings 5–10 cm (2 to 4 in) long planted in a cold frame in early autumn. Several species may be raised from seed sown under glass in late winter.

Griselinia Readily increased by semi–ripe cuttings in a frame or case in summer. Also from seed in a greenhouse in spring.

Hamamelis In nurseries, grafting under glass in late summer is the usual method of increase. The stock used is *H. virginiana* (witch hazel) which is raised from seed sown under glass. Layering in spring is the most suitable method for the amateur.

Hedera (ivy) Ivy is normally increased from cuttings, either the softwood type taken in summer and inserted in a close case or frame, or firm nodal cuttings planted in a cold frame in mid-autumn.

Hibiscus Seed of the annual varieties is sown outdoors in spring. Shrubby types are increased by semi–ripe cuttings in a greenhouse in mid- or late summer, or by layering outdoors earlier in the summer. Some varieties are grafted in a greenhouse in spring.

Hippophae (sea buckthorn) Increased by suckers outdoors in spring, by autumn-sown seed and by layering.

✳ **Hydrangea** *H. macrophylla*, (syn. *H. hortensis*), and its varieties are best ✳ propagated from semi–ripe cuttings taken from plants grown indoors or outdoors. *H. petiolaris* is raised from seed sown in heat in spring or by serpentine layering.

Hypericum The species are raised from seed stored dry and sown in spring under glass. Cuttings are also generally used, either of firm wood with a heel inserted in a cold frame in autumn or softwood cuttings in a close case in summer. *H. calycinum* (rose of Sharon) is easily increased by division.

Ilex (holly) The common holly and other species are raised from seed stratified for 18 months and sown outside in spring. Layering in the autumn is another method. For this purpose, stock plants must be partly lifted and laid on their sides, and each young shoot should be tongued before being layered. Budding of varieties on the type plant is also done.

Jasminum All species and varieties are raised from cuttings or serpentine layering. Hardwood cuttings with a heel are taken in late autumn and inserted in a cold frame, or in a sheltered position in the open. Semi-ripe cuttings also strike readily in a frame in summer.

Kalmia Seed is sown in peaty compost under glass early in the spring. Layering is the best method for increasing varieties. It should be done in the autumn, the layered shoots being twisted or tongued to wound them and accelerate rooting. *K. polifolia* will grow from semi-mature cuttings taken in late summer and inserted under bell glasses.

Kerria (Jew's mallow) Cuttings of firm shoots with a heel root when planted in the open in autumn. Softwood tip cuttings taken in summer will also strike in a close case.

Kolkwitzia amabilis Take semi-hardwood cuttings in summer and insert in a close case.

Laburnum *L. anagyroides* (common laburnum, golden chain) is easily raised from seed sown in the open in spring. Hardwood cutting 23–30 cm (9–12 in) long with a heel are used for all varieties. Varieties may also be budded or grafted on common laburnum.

Laurus *L. nobilis* (bay laurel) is normally raised from semi-ripe cuttings of firm shoots with a heel. Layering is another method sometimes used.

Lavatera (tree mallow) Shrubby species are increased by seed in gentle heat in spring. Summer cuttings will also strike, and plants can be divided.

Leycesteria Easily increased from summer cuttings in a shaded frame, also by spring-sown seed.

Ligustrum The ordinary privet *L. vulgare* is very easily propagated from hardwood cuttings which are made 25–30 cm (10–12 in) long and inserted in the open in autumn and winter. Various other species and varieties are best increased from softwood cuttings about 8 cm (3 in) long with a heel, and planted in a close case.

Lippia (sweet scented verbena) Shrubby species are easily increased by summer cuttings.

Lonicera The climbing types are increased by serpentine layering, by semi-ripe cuttings about 15 cm (6 in) long inserted in a cold frame or by softwood tips 8 cm (3 in) long, with or without a heel, inserted in early summer in a close case. The shrubby species are increased by semi-ripe or hardwood cuttings.

Magnolia Layering is probably the best method and should be done in spring. Only young shoots should be layered after being slit to form a tongue. Many species can be raised from seed which should be sown in the open when ripe. Small quantities may be sown in boxes or pans, left outside until late winter and then brought indoors. Varieties are grafted under glass on seedling stocks.

Mahonia Raised from seed in the same manner as berberis. The valuable species, *M. bealei*, may be propagated from leaf bud cuttings taken in summer. These are made about 15 cm (6 in) long and inserted in a close case.

Malus (apple) The flowering apples are usually grafted on to seedlings of *M. pumila* (wild crab).

Olearia The species and varieties are readily increased from semi-ripe shoots in summer in a close case or frame.

Osmanthus Cuttings of semi-ripe shoots with a heel taken in the autumn will strike in a cold frame. Most of the species are, however, layered, the young shoots being well tongued before pegging down.

Osmarea Increased by semi-ripe cuttings.

Passiflora (passion flower) Increase is by seed, summer cuttings or serpentine layering.

Pernettya Variable plants are produced from seed sown in the spring in the open. Varieties are raised from cuttings of small sideshoots taken in summer and inserted in a close case. Division is an easy method for the amateur and should be done in spring. Young shoots may be wounded and layered in autumn.

Perowskia This is increased from semi-mature shoots without a heel secured in summer and inserted in a close case.

Philadelphus All the species are easily increased from cuttings of ripe shoots about 23 cm (9 in) long with a heel planted in the open in a sheltered position in a well-drained light soil. Softwood sideshoots 8–10 cm (3–4 in) long taken in mid-summer will strike in a frame or close case.

Phlomis *P. fruticosa* (Jerusalem sage) and other species will root from cuttings of ripe shoots about 8 cm (3 in) long when planted in a cold frame in the autumn. Softwood nodal cuttings also strike readily in a close case. All may be raised from seed.

Pieris Cuttings of semi-ripe sideshoots about 8 cm (3 in) long with a heel may be rooted in a frame in late summer or early autumn. Autumn layering is another method.

Piptanthus Sow seed in boxes or pans in heat in spring.

Pittosporum The species are raised from seed sown under glass in early spring in a peaty compost. Cuttings of semi-ripe shoots with a slight heel will root in a close case in summer. Varieties are grafted on to seedlings of the parent species under glass in winter.

Polygonum The popular climber *P. baldschuanicum* is propagated from hardwood cuttings 15–20 cm (6–8 in) long with a heel. These should be potted singly into 8 cm (3 in) pots which are stood on a glasshouse staging or cold frame.

Populus (poplar) Hardwood cuttings about 20 cm (8 cm) long, planted in a light sandy soil in the autumn, strike readily. Some species can be increased from suckers or root cuttings.

Potentilla The shrubby species such as *P. fruticosa* may be raised in summer from small softwood sideshoots with a heel inserted in a close case.

Prunus All these are budded or grafted on to various stocks in a similar manner to fruit trees. Budding is usually done in the summer and grafting in mid-spring in the open on established stocks. The stocks used include *P. avium* (sweet cherry) which is preferred for the Japanese cherries; *P. cerasifera* (myrobalan) is used for the varieties of this species:

Streptocarpus are easy to increase by leaf cuttings.

seedling peach for the flowering peaches and Brompton or St Julian for *P. salicina* (syn. *P. triloba*) and its varieties.

Pyracantha (firethorn) The species and varieties are increased by inserting small semi-mature sideshoots with a heel in a cold frame in late summer, or by seed.

Pyrus (pear) To produce pear seedlings, the pips should be stratified over winter and the seed sown in spring in the open or in boxes in a cold frame or glasshouse. Other species and varieties may be grafted on to these seedlings in the open in early spring.

Rhododendron This very large genus is increased by various methods. Seed is used to raise *R. ponticum* (common rhododendron), which is widely used as a stock for grafting. The seed is collected when ripe and sown in the open in spring or in pans or boxes in a cold frame. A wide range of species are also increased by seed. This requires great care in sowing, which is done in pots or pans containing sifted peat. In nurseries, a large number of varieties are grafted, this being done in winter under glass. Layering is also widely used in nurseries, stock plants being established in peaty soil. Young shoots are layered in autumn and before each is pegged down should be wounded. Cuttings are used mainly for the smaller species such as *R. impeditum* and *R. racemosum*, and also for the evergreen azalea section. Semi-mature sideshoots are taken in summer and inserted in pots or pans, using a compost of peat and sand. The receptacles are placed in a close case.

Rhus *R. typhina* (sumach) is increased by root cuttings about 4 cm (1½ in) long which are secured in winter and inserted singly in small pots. *R. cotinus* (smoke tree) comes true from seed which is sown in the open in autumn. Varieties of this species are mound-layered in autumn. (see Blackcurrants, page 71).

Ribes The flowering currants are easily propagated from hardwood cuttings about 15 cm (6 in) long inserted in the open in the autumn.

Romneya (California tree poppy) The species, *R. coulteri* and *R. trichocalyx*, are propagated from root cuttings which are made into lengths of about 2 cm (1 in) and are inserted horizontally in pots or boxes of sandy compost in winter.

Rosmarinus (rosemary) Readily increased from cuttings in late summer, inserted in a cold frame.

Rubus The ornamental brambles are increased by division or by tip-layering. *R. cockburnianus* (syn. *R. giraldianus*) is increased by root cuttings 4 cm (1½ in) long, which are inserted singly in pots placed under glass.

Ruscus (butcher's broom) Propagate by dividing the creeping root stock in spring.

Salix (willow) One of the easiest plants to increase from almost any type of stem cutting.

Salvia (sage) The shrubby species are easily increased from softwood cuttings in a close case.

Santolina chamaecyparissus (lavender cotton) Softwood side-shoots about 8 cm (3 in) long taken in summer root readily in a close case.

Sarcococca Easily increased by division, semi-ripe cuttings and seed in spring.

Schizophragma (climbing hydrangea) Usually increased from cuttings with a heel in summer in a close frame.

Senecio The shrubby species such as *S. greyii* and *S. laxifolius* root from cuttings of semi-mature sideshoots in a close frame in late summer.

Skimmia Ripened sideshoots taken in the autumn strike in a cold frame but are rather slow. Layering in the autumn is another method.

Solanum The shrubby climbing species such as *S. jasminoides* are readily increased from soft or semi-ripe cuttings in early summer.

Spartium *S. junceum* (Spanish broom) is easily increased from seed sown in spring under glass.

Spiraea All the spiraeas can be propagated from cuttings either of the hardwood type about 20 cm (8 in) long taken in the autumn with or without a heel and planted in the open, or softwood cuttings stuck in a close case or frame. Division is used for the dwarf stooling kinds such as *S. douglasii* and *S. japonica* and its varieties.

Symphoricarpos (snowberry, St Peter's wort) Easily increased by hardwood cuttings taken in mid-autumn and planted in the open or by division.

Syringa (lilac) Tongued young shoots are layered in the spring. Varieties are also grafted on seedlings of *S. vulgaris* (common lilac). This is done under glass in the spring. Privet, which is sometimes used for grafting on, is a bad stock. Some varieties can be rooted from half-ripened cuttings inserted in a cold frame in the summer.

Tamarix Cuttings made from hardwood 20–25 cm (8–10 in) long are planted in the open in sheltered positions in autumn.

Teucrium Increase by division or semi-ripe cuttings. The shrubby species root from small soft sideshoots taken in summer and planted in a frame.

Ulex (gorse) Stratified seed sown outside in spring or in pots placed in a cold frame germinates freely. Sideshoots taken in mid-summer are inserted in a frame and kept syringed frequently.

Veronica Semi-mature sideshoots about 8 cm (3 in) long with a heel root readily in a cold frame in the summer.

Viburnum Several species are increased from seed sown in pots in spring and placed in a cold frame or glasshouse. Layering is another method used for *V. opulus* (guelder rose) and its varieties. Semi-mature shoots of certain species such as *V.* × burkwoodii, V. carlesii, V. fragrans and *V. tinus* (laurustinus) strike readily in a close case in summer.

Vinca (periwinkle) Division in early spring is a simple method. Soft tips root easily in a close case in summer and cuttings of ripe shoots can be struck in a cold frame in the autumn. Serpentine layering is also successful.

Vitis (vine) Varieties of the grape vine and *V. coignetiae* can be increased from vine eye cuttings in a greenhouse in winter, or by serpentine layering.

Wisteria *W. sinensis* can be raised from seed sown in heat in early spring. Varieties are root grafted on to these seedlings when established in winter. Young shoots may be layered into pots in the autumn.

Yucca Root cuttings 5–8 cm (2–3 in) long are secured in winter and inserted in boxes of sandy soil which are placed in heat.

CHAPTER 15

FRUIT

BERRY FRUITS

All the berried fruits (also known as soft fruits or bush fruits) are grown on their own roots and as a general rule are easy to propagate.

Blackcurrants *Ribes nigrum* The usual method of increase is by hardwood cuttings which should preferably be secured immediately after leaf-fall. It is important to take cuttings only from healthy, heavy-cropping bushes and to avoid those that have shown symptoms of 'big bud' or virus in the leaves.

Select firm, well-ripened shoots of the current season's growth, but if these are hard to find, 2- or 3-year-old wood may be used. Cut the shoots about 25 cm (10 in) long at a leaf joint. Select a well-drained site and insert the cuttings about 15 cm (6 in) deep in the soil. When growth commences the following spring, the topmost buds produce shoots which arise at or just below the soil surface, forming a 'stool' type of bush without a main stem or 'leg'.

In the autumn, the young plants may be moved to a nursery bed about 45 cm (18 in) apart or to permanent positions at 90 cm (3 ft) spacings. After transplanting, cut back the young growths to within 1 or 2 buds of their point of origin.

Blackcurrants can also be easily propagated by mound layering. The method is to cut a plant intended for propagation back to near ground level. The young shoots that arise from the base are kept mounded up as they grow and will produce roots below the soil. These rooted shoots may be pulled off and transplanted in the autumn.

Red and whitecurrants *Ribes sp.* Normally these are also increased from hardwood cuttings, firm 1-year-old stems being preferable. Take cuttings 25–30 cm (10–12 in) long in the autumn and cut off all buds with a sharp knife except the top 3 or 4. This means that shoots will be produced only from near the tip of the cutting, to form eventually the head of the bush. Unlike blackcurrants, the red and white types are best grown with a clean stem or leg.

Gooseberries *Ribes grossularia* The method is the same as for red and whitecurrants, making a bush with a clean stem or leg.

Raspberries *Rubus idaeus* These may be said to propagate themselves with very little assistance from the gardener. Often, indeed, the number of new canes that arise naturally alongside the permanent row are an embarrassment. These young canes may be dug up and transplanted in the autumn or winter and afterwards cut down to within about 23 cm (9 in) of ground level. Raspberries may also be increased from root cuttings.

Blackberries and loganberries *Rubus sp.* These and related fruits are normally increased from tip layers, and this may be done in summer. Light soil well supplied with organic matter is most suitable. Make a slit in the soil with a spade or trowel and insert the tips of young canes in this to a depth of about 15 cm (6 in). The soil is then trodden firm. New plants can be severed from their 'parents' and transplanted the following spring.

Strawberries *Fragaria* Runners are the natural and speedy method of increase. After the fruit has been picked and the bed weeded, the runners may be pegged down to enable them to root more quickly. These may be transplanted from midsummer to autumn, but the earlier the better as runners planted in summer often produce a number of fruits the following season, whereas this is unlikely with later planting.

TREE FRUITS

The techniques of budding and grafting described earlier are the methods used for 'making' fruit trees, but they demand a fair degree of expertise and preparation and are generally left to a skilled nurseryman. A scion, which eventually becomes the tree head, is budded or grafted to a rootstock which supplies the root system. The rootstock has considerable influence on the rest of the tree, particularly during the early years, affecting rate of growth, height and spread and the age when the tree commences fruiting.

Fruit growers use grafting techniques to change completely the variety of an apple or pear tree, and nurserymen are able to supply trees on which different branches produce different varieties of the same fruit.

CHAPTER 16

RAISING VEGETABLES AND HERBS

Most vegetable crops are produced from seed but a few of the perennial kinds, such as rhubarb and seakale, are propagated by vegetative means. Seed is sown in drills which facilitates weed control. The following alphabetical list shows how the different kinds are increased.

Artichoke, globe Usually increased from suckers which are found at the base of established plants. These are simply cut off with a sharp knife and planted in the open 10 cm (4 in) deep and about 60 cm (2 ft) apart.

Artichoke, Jerusalem These are very hardy and increase naturally by tubers. Plant any time from late winter to mid-spring in rows 75 cm (30 in) apart, allowing 30 cm (12 in) between the tubers.

Asparagus Asparagus can be raised from seed, which may be sown quite early in the spring in drills 45 cm (18 in) apart. After covering with soil the bed should be rolled. The seed is slow to germinate. The seedlings are thinned out to 8 cm (3 in) apart leaving the strongest. During the summer, hoe frequently, and watering in dry weather is beneficial. The following spring the plants may be moved to their permanent site but it will be three years before they can be cropped.

Basil, sweet This is an annual herb which is raised by sowing seed in heat in mid-spring and planting out in early summer.

Beans, broad For early crops, seed of the long-pod varieties are sown in late autumn, except in cold districts. A sowing may also be made in a warm border in late winter. Another method is to start the plants in a cold frame and transplant them later. Alternatively, early sowings may be protected by covering the rows with cloches for a period. Beans are usually sown in double rows, the seed being spaced 23 cm (9 in) apart each way. The double rows should be 75 cm (30 in) apart.

Beans, kidney or French Sow in the open in latish spring in drills

5 cm (2 in) deep and 45 cm (18 in) apart. Space the seeds 8 cm (3 in) apart and thin out to 15 cm (6 in) apart. Early crops are secured by sowing in early spring in cold frames in drills across the frame 30 cm (12 in) apart. When the plants are through, cover the frames with mats on frosty nights. Double rows sown a little later may be given protection for a period with cloches.

Beans, runner Sow in the open in late spring. Earlier sowing may be done in glasshouses or frames for transplanting but plants must not be set out until after the last frost. For training on sticks or poles the seeds are spaced 12 cm (4½ in) apart in double rows 23 cm (9 in) apart, the double rows being not less than 1.8 m (6 ft) apart. The plants are usually thinned out to leave 23 cm (9 in) between them.

Beetroot Sow in late spring in rows 30 cm (12 in) apart. Thin out the plants to about 5–8 cm (2–3 in) apart. Wider spacing usually results in poorer quality and lower yield.

Borage This annual herb may be sown in spring and thinned out to 30–40 cm (12–16 in) apart.

Broccoli Sow according to variety from mid- to late spring in drills 30 cm (12 in) apart. Plant out about 6 or 8 weeks later. Space the plants about 70 cm (27 in) apart each way.

Brussels sprouts For early crops the seeds may be sown in a glasshouse or cold frame, the plants being set out in mid- to late spring. In some districts autumn sowings are made in sheltered borders for spring transplanting; seed for the main crop is usually sown in drills in early spring. These are transplanted in early summer. Set the plants 90 cm (3 ft) apart each way.

Cabbage This vegetable can be harvested for much of the year with careful planning. The earliest varieties can be sown in trays in a warm greenhouse late in the winter and pricked out into pots when they are large enough to handle. Harden them off in a cold frame before planting out in spring; they will mature in early summer.

The same varieties can be sown out of doors later in the spring, along with several others, to mature later in the summer. Sow thinly in a seedbed, transplanting the seedlings when they are about 8 cm (3 in) tall.

Spring-maturing cabbages are sown in a seedbed in summer and transplanted after summer crops are cleared. They stand over winter and can be harvested from early spring.

Variety	Sowing time	Harvest
January King	early – late spring	winter
Holland Late Winter	early – late spring	late autumn – early spring
Harbinger	summer	mid-spring
Wheelers Imperial	summer	spring
Hispi	winter (greenhouse)	late spring
Golden Acre May Express	late winter (greenhouse)	early summer
June Star	late winter (greenhouse)	early summer
Greyhound	early spring	summer
Winnigstadt	early spring	summer
Xmas Drumhead Early	spring	autumn

Carrots For early crops, the seed may be sown broadcast thinly in a cold frame in autumn or mid-winter and left there until ready for pulling. The stump-rooted type may be sown in a warm border in early spring. Main crop carrots are usually sown in mid- to late spring. Make the drills 1–2 cm (½–1 in) deep and 30–45 cm (12–18 in) apart. When sown evenly and not too thickly there is no need to thin out.

Cauliflower Early cauliflowers can be obtained by sowing in mid-autumn, overwintering the plants in pots in a cold frame, and planting them out in spring to crop in early summer. Alternatively they can be sown under glass in winter to be planted out in the spring. Suitable varieties for this treatment are Snowball, Andes and Alpha Polaris.

The majority of varieties are sown outdoors in a seedbed in late spring. It is important to select varieties carefully: Flora Blanca and Autumn Giant will mature during autumn but must not be sown earlier than late spring or they will not form good heads. All The Year Round can be sown any time in the spring, and Australian types such as Wallaby and Barrier Reef are best sown fairly late in the spring. These will all mature in very late summer or autumn.

Winter cauliflowers (which used to be known as heading broccoli) are also sown in a seedbed in late spring. In mild areas early varieties such as Angers No. 1 Superb Early White can be grown to mature through the latter part of the winter. In other areas the hardier varieties such as English Winter, Northern Star and Walcheren Winter will be usable at the end of spring.

Celeriac or turnip-rooted celery Sow quite early in the spring and prick off into boxes or a cold frame. Harden off and plant out in a 30 cm (12 in) square in very early summer.

Celery Sow in a heated glass house in late winter or early spring and treat similarly to celeriac. Once hardened off, plant out in trenches (or in beds for self-blanching celery). Seed may also be sown in a cold frame in early spring. Pelleted seed can be used.

Chicory Sow in very late spring in rows 30 cm (12 in) apart. Thin the plants out to 15–20 cm (6–9 in) apart. Lift the roots in mid-autumn and trim off the leaves, cutting them about 1 cm (½ in) above the crown. Put them in a box covered by moist peat or soil in complete darkness, and keep them at a steady temperature of around 10° C (50° F) to produce chicons for salads.

Chives Lift and divide the clumps every three or four years in the spring. May also be grown from seed sown in spring.

Endive This autumn and winter salad crop should be sown for succession throughout the summer. Spacing is similar to lettuce.

Horseradish Easily propagated from root cuttings about 8 cm (3 in) long. These are inserted in holes made with a long dibber and afterwards filled in with soil. Plant in spring 45 × 30 cm (18 × 12 in) apart.

Kale Sow thinly in a seedbed in spring in rows 30 cm (12 in) apart. Transplant when large enough, leaving 60–75 cm (2–2½ ft) between plants, depending on variety.

Leeks The seed is usually sown quite early, in shallow drills 40 cm (15 in) apart. When large enough to handle, the seedlings are transplanted 15 cm (6 in) apart in deep drills made about 30 cm (12 in) apart. To secure plants earlier a sowing may be made in a cold frame in late winter.

Lettuce The majority of varieties are sown in drills outdoors all through the spring for successional cropping. They should be ready to cut within 70–100 days of sowing, depending on variety and weather conditions.
Certain varieties such as Valdor and Winter Density can be sown

outdoors in early autumn and left to stand over winter; protection with cloches is helpful but not essential. They will mature in late spring.

Greenhouse varieties such as Kwiek, Dandie and Marmer are sown from late summer through the autumn in the border of an unheated or slightly heated greenhouse. They will be usable from the onset of winter until mid-spring.

With the aid of a greenhouse it is therefore possible to harvest lettuce virtually throughout the year as shown below.

Variety	Sowing time	Harvest
Dandie	early autumn (greenhouse)	mid- to late winter (greenhouse)
Marmer	autumn (greenhouse)	early spring (greenhouse)
Valdor	early autumn (outdoors)	late spring
Fortune	late winter (greenhouse, transplant outside)	early summer
summer varieties	early spring–early summer	summer- mid-autumn
Kwiek	late summer (greenhouse)	winter (greenhouse)

Marjoram pot, A perennial herb which is increased by sowing seed in spring. Plant about 23 cm (9 in) apart.

Marjoram, sweet This herb is treated as an annual and is usually sown in heat in spring. Transplant in late spring 23 cm (9 in) apart.

Mint Easily propagated by division of the underground stems in spring or autumn. Summer cuttings also root in a shady situation in the open.

Onions Bulb onions are produced by sowing in the open in early spring. Earlier sowing may be made in heat and the plants set out when large enough to handle. A sowing can also be done in late summer and the seedlings transplanted the following spring. The variety White Lisbon is often sown then for use as a salad onion in spring. Bulb onions are grown in rows 30–38 cm (12–15 in) apart, the plants being spaced about 10 cm (4 in) apart. Pelleted seed is available for space-sowing. Shallots are increased by division of the bulbs which are lifted in the autumn and planted in spring 15–20 cm (6–8 in) apart.

1

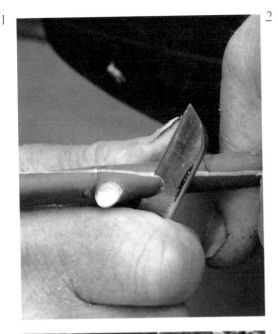

2

5

Air layering: Ficus:
(*1*) Making cut in stem.

(*2*) Stuffing cut with moss.

(*3*) Wrapping plastic around moss.

(*4*) The moss fully wrapped and plastic secured by tape.

(*5*) Cutting the rooted layer from the parent plant.

(*6*) Cutting off stump of stem immediately below the new roots.

Parsley Sow in early spring for summer cutting, in early summer for winter produce, and again in late summer for use the following spring. The seed is slow to germinate. Drills should be spaced 30 cm (12 in) apart and the plants thinned to 8–10 cm (3–4 in) apart.

Parsnips Sow from late winter to mid-spring in drills 40–45 cm (16–18 in) apart. Thin out plants to 15 cm (6 in) apart.

Peas In mild districts, early peas may be sown in late autumn. Normally successional sowings are made from early spring until early summer. Early sowings may be protected by cloches. Use early, mid-season and later varieties according to season, but early (quick-maturing) varieties should be chosen for late sowing. Ordinary V-shaped drills about 6–8 cm (2½–3 in) deep, and spaced about the same distance apart as the plants are expected to grow, are suitable.

Radishes Successional sowings can be made from the end of winter until late summer in the open. Earlier sowings may be protected in cold frames or under cloches. Seed is usually sown broadcast or in drills 15 cm (6 in) apart.

Rhubarb Divide the old roots in autumn or spring. Each portion should have at least one crown or 'eye' and is called a set. The sets are usually planted in rows 90 cm (3 ft) apart with 75 cm (2½ ft) being allowed between the sets. Rhubarb can also be raised from seed. Sow in early spring in drills 2 cm (1 in) deep and 30 cm (12 in) apart. Thin the plants to 23 cm (9 in) apart, and transplant in the autumn or spring.

Sage Easily increased from seed sown outdoors in spring. Soft cuttings root readily in spring or early summer and plants so raised are often preferable to those from seed. Plant 38 cm (15 in) apart.

Salsify Sow in drills 2 cm (1 in) deep and 30 cm (12 in) apart in spring.

Scorzonera Treat in the same manner as salsify.

Seakale Usually increased from root cuttings or thongs about 50 mm (¼ in) thick and 15 cm (6 in) long. These are prepared in the autumn, tied in bundles and laid in sand. The sprouts which appear on each cutting in spring should be rubbed off, except the strongest. The thongs are then planted about 2 cm (1 in) below the surface in rows 45 cm (18 in) apart and 38 cm (15 in) between the cuttings. Seakale may also be raised from seed sown in the open in early spring.

Spinach A succession of summer spinach is maintained by sowing at two- to three-week intervals from early spring until mid-summer. Winter spinach is provided by making one sowing towards the end of summer and another in early autumn. Sow in drills 2 cm (1 in) deep and 30 cm (12 in) apart, and thin to 15 cm (6 in).

Spinach beet Two sowings are made for succesion, one in spring and one in mid-summer. Sow in drills 38 cm (15 in) apart and thin out the plants to 15–20 cm (6–8 in) apart.

Thyme The common thyme is easily raised from seed or by division in spring. Plant about 10 cm (4 in) apart.

Turnips Turnips and swedes are raised by sowing seed throughout spring. Earlier sowings may be made in frames or protected by cloches. The plants are usually thinned out to about 10 cm (4 in) apart.

HALF-HARDY/TENDER VEGETABLES

Tomatoes This is one of the most popular and valuable crops for the garden and greenhouse. Outdoor varieties cannot safely be planted until risk of frost is past, and all should be raised from seed in a heated greenhouse enjoying good light. Use a reliable seed sowing compost and space seeds about 1 cm (½ in) apart in trays or pots covered with about 25 mm (⅛ in) of the same compost. During the night, maintain a temperature of 18° C (65° F), which may rise to over 21° C (70° F) during the day. When the seed leaves have expanded, usually eight to ten days after sowing, set the seedlings singly in 12 cm (4¼ in) plastic pots, using the potting mix of your choice. From that time the night temperature is lowered to 15° C (60° F). Water the young plants frequently in sunny weather, less frequently when dull.

When the seedlings are well established in their pots, they need to be fed with a liquid fertilizer formulated for tomatoes. Space out the plants as they grow so that the leaves do not overlap.

Planting out in the greenhouse border or into larger pots is best done when about half the plants have their first flower truss showing. Outdoor varieties should be accustomed to lower temperatures once established in their pots and can be hardened off in a cold frame during the latter part of spring.

Cucumbers Propagation of cucumbers is somewhat less exacting than tomatoes. The best method is to sow the seed singly in 8 cm (3 in) pots, using a potting compost rather than a sowing compost in this case, as the cucumber is a gross feeder. Fill each pot and press a seed into the centre 1 cm (½ in) deep. Give a good watering, then place glass and paper over the pots.

Maintain a temperature of about 21° C (70° F) and a moist atmosphere, and germination occurs in 36 to 48 hours. Then remove the glass and paper but avoid exposing the seedlings to strong sunlight. Continue to grow in a temperature of 18–21° C (65–70° F) and water when necessary.

When the roots have reached the sides of the pots, pot on into 15 cm (6 in) pots using a similar or richer compost, and once established, apply liquid feeds with some waterings. Support the plants as they grow with 60 cm (2 ft) canes inserted in the pots. Plant out in the greenhouse beds before there is any risk of a check to growth in the pots.

Cucumbers for outdoor culture – ridge cucumbers – do not possess the fine quality of the greenhouse type, but are raised from seed in the same way, given gradually cooler conditions, hardened off and planted in rich soil when the risk of frost is past.

Melons Melons may be raised in the same manner as cucumbers except that the mixes used need not be so rich. Apart from greenhouse cultivation, melons are suited to being grown in frames or under cloches if planted out after spring frost.

Vegetable marrows and squashes Sow seed singly in 8 cm (3 in) pots in mid-spring. Germinate, and grow in heat for a period. Harden off, and plant out towards the end of spring. The seed may also be sown a little later in a cold frame or in the open in late spring.

CHAPTER 17

INCREASING DECORATIVE GLASSHOUSE AND ROOM PLANTS

Under this heading a wide range of ornamental foliage and flowering plants is included. Most of them are increased by seed or cuttings involving the general principles outlined in previous chapters in relation to these methods. Thus seeds are usually sown in sowing compost and potted into potting compost with extra feeding according to the type of plant and the length of time it is likely to remain undisturbed, after potting.

The seeds of glasshouse plants are usually germinated in a temperature of 13–18° C (55–65° F), and a similar temperature is allowed for the propagation of cuttings.

Abutilon (Indian mallow) Sow seeds in a greenhouse any time from mid-winter to mid-spring, or take semi-ripe cuttings in warm conditions during the summer.

Acacia (wattle) Insert semi-ripe cuttings with a heel taken in summer in pots containing a peaty compost. Plunge the pots in a close case; when rooted, pot off singly and grow in cool conditions. Seeds sown when ripe in a mixture of peat and sand germinate freely in a temperature of 15° C (60° F).

Acalypha (copper leaf) Cuttings should be struck in brisk heat in spring.

Achimenes Sow seed in early spring with care, as it is very small. Germinate in a temperature of 15–18° C (60–65° F). Prick off the seedlings in light peaty compost and grow in a similar temperature. Cuttings secured from plants started in heat should be inserted in a close case, using a compost of peat and sand with bottom heat. Leaf cuttings also root under similar conditions when their stalks are inserted in the compost.

Aechmea Readily increased from suckers which develop naturally. These should be potted and kept in a close warm atmosphere until rooted.

Agapanthus (African lily) Divide the plants just as growth begins in the spring, and pot the rooted portions.

Agave (American aloe) Offsets which are freely produced are a simple means of increase.

Anthurium Divide the roots in early spring when re-potting. Sow seed in a mixture of chopped sphagnum moss, charcoal and sand, temperature 21–26° C (70–80° F).

Aphelandra Increase from cuttings of semi-ripe wood or softwood cuttings in bottom heat of 21° C (70° F).

Aralia Soft cuttings made from sideshoots can be rooted in a close case. Root cuttings taken in spring will also grow. These are made about 5 cm (2 in) long and are inserted in pots of sandy soil. The pots are placed in a close case and bottom heat promotes rooting. Varieties and species can be grafted under glass on the stock *A. reticulata*.

Araucaria excelsa (Norfolk Island pine) Seed in heat is the usual method.

Ardisia (spear flower) Seed may be extracted from ripe berries and sown immediately in heat. Cuttings taken during spring and summer will strike in heat.

Asparagus The well-known asparagus ferns *A. plumosus* and *A. sprengeri* are raised from seed. Sow in spring in sowing compost. Pot singly and grow in a fairly moist warm atmosphere. Roots may also be divided when re-potting large plants in spring.

Aspidistra Divide and re-pot during the growing season. Shade for a few days afterwards.

Auricula *Primula sp.* Named varieties are increased from rooted offsets secured when re-potting after flowering. These are inserted in small pots. Unrooted offsets will strike when inserted in a close case. Also raised from seed sown in spring.

Begonia Most of the begonias may be increased from seed, which is sown in early spring. The seed is very small and requires a fine surface and little or no covering. Germinate in a temperature of 15–18° C (60–65° F). The ornamental-leaved kinds such as *B. imperialis* and *B. rex* are increased by leaf cuttings. The fibrous-rooted kinds, like *B. socotrana* and its varieties such as John Heal and Gloire de Lorraine, are readily increased from stem cuttings secured from plants started in heat in early spring. The propagation of tuberous-rooted begonias is mainly by division of the tubers.

Beloperone Easily increased from soft or semi-ripe cuttings in moderate bottom heat.

Bignonia (cross vine) These climbing plants are increased by soft cuttings in spring inserted in a close case. Young shoots may also be layered in the late summer.

Billbergia Increased by division, the suckers being pulled carefully apart. Pot separately.

Bougainvillea Increase by semi-ripe cuttings in a greenhouse during the latter part of the summer.

Bouvardia Increase is by soft cuttings about 5 cm (2 in) long taken in mid- to late spring and inserted in pots of sandy compost. Plunge the pots in bottom heat in a case. Root cuttings about 2 cm (1 in) long will grow when planted in pans in spring.

Browallia Raised from seed sown in spring or summer for late winter and spring flowering.

Brunfelsia Cuttings root readily in moderate heat. Seed should be germinated in a temperature of at least 21° C (70° F).

Cacti and succulent plants This large and variable group of interesting plants is not difficult to propagate. Seed and cuttings are the principal methods. Seeds are usually obtained from the plant's native country, but are sometimes home-saved. Sowing may be done at any time if heat is available, but spring sowing is generally preferable. Sow thinly in well-drained pans of fine compost and cover according to the seed size. Plunge the pans in peat or ashes and cover them with paper to prevent rapid drying. Maintain a temperature of 10–13° C (50–55° F) and remove the paper immediately germination is observed. Prick off the seedlings in light compost as soon as they are large enough to handle.

Generally speaking cacti and succulents are very easy to root from cuttings but some species such as the spherical kinds, do not provide any material for this purpose. Any branching plant however will usually provide cuttings and offsets can be secured from others. Usually the cuttings should be dried for a few days before insertion. A compost of peat and sand is quite suitable but pure sand also serves the purpose. There is no need to provide close conditions but slight shade may be beneficial. Cuttings of strong-growing species such as the opuntias may be inserted singly in pots. Many kinds can be raised from leaf cuttings. Grafting is sometimes used for a few special kinds.

Propagation of camellia: dipping cutting in hormone powder prior to insertion in pot of compost.

Caladium Propagated by division of the tubers in early spring.

Calathea (zebra plant) Easily increased by division.

Calceolaria (slipper flower) Raised from seed sown in the greenhouse in summer. Prick off and pot on as required.

Callistemon (bottlebrush tree) Take cuttings of firm young shoots in summer and insert in pots containing a compost of peat and sand.

Campanula isophylla This species which is popular for hanging baskets should be raised annually from cuttings taken in summer.

Canna (Indian shot) Sow seeds singly in small pots 2–5 cm (1–2 in) deep after soaking in water for 24 hours. Sowing is done in very early spring, the pots being kept in a temperature of 21° C (70° F). Named varieties are raised by dividing the roots at potting time.

Carnation Perpetual flowering carnations are propagated by cuttings which should be selected with care. The best are those made from sideshoots found about the centre of the flowering stems. Cut below a node, remove the lower leaves and trim off the leaf tips. Insert the cuttings in clean sharp sand and maintain a temperature of 10–13° C (50–55° F). Bottom heat is also beneficial. Water after insertion, and subsequently keep the medium nicely moist. Keep the cuttings close at first, but free ventilation should be given when rooting commences in a few weeks' time. Rooted cuttings are potted into 8 cm (3 in) pots using a gritty compost. Carnation cuttings of this type are normally taken from mid-winter to early spring but they may be struck at other seasons.

Malmaison carnations are best increased by layering in early summer. The rooted layers are potted in late summer.

Cassia (senna plant) Cuttings of semi-ripe wood will strike. Seed is also used.

Celosia (cockscomb) This annual makes an attractive pot plant. For this purpose the seed is sown in a warm house in early spring and is first pricked off and then potted on as necessary.

Chlorophytum (spider plant) Readily increased by division and runners.

Cineraria Raised from seed, which is sown for succession during spring and early summer. Pot the seedlings into mix and later into the same compost in larger pots.

Cissus Propagated by softwood cuttings about 5 cm (2 in) long with a heel.

Clivia Sow seeds in a warm house in spring and keep the seedlings growing until they flower. Division of old plants can be done when re-potting at the end of the winter.

Cobaea scandens (cup and saucer plant) Easily raised from seed sown in a warm house in early spring. The seedlings flower the same year. The variegated type is raised from cuttings of firm sideshoots taken in high summer and inserted in a close case.

Codiaeum (croton or South Sea laurel) Easily increased by stem cuttings, which are inserted singly into pots. The latter are plunged in peat in a warm frame with bottom heat.

Coleus (flame nettle) Plants of variable types may be raised from seed sown in gentle heat in early spring. Cuttings of young shoots can be rooted at any period in sandy compost and kept close. Cuttings are essential to increase special forms.

Columnea Cuttings from firm shoots will strike in a close warm frame.

Cordyline The stove and glasshouse species are increased by cutting the main stem into pieces 2–5 cm (1–2 in) long. These are inserted in a close warm case in sandy soil. The half-hardy species such as *C. australis* are easily raised from seed.

Crassula Sow seed in a warm house in early spring and keep the seedlings rather dry. Stem or leaf cuttings strike readily in pots of sandy compost stood on a glasshouse staging.

Crossandra Easily increased by cuttings with good bottom heat throughout the year.

Cryptanthus Offsets should be separated and potted up singly in the spring in brisk heat.

Cuphea (Mexican cigar flower) Increased from seed sown in early spring in gentle heat. Cuttings of young shoots can be struck in spring or summer.

Cyclamen Propagated by sowing seed in late summer or mid-winter. Sow each seed 50 mm (¼ in) deep and 2 cm (1 in) apart in pans. Allow a temperature of 10–13° C (50–55° F). Shade the seedlings from bright sun and pot when large enough to handle. Large tubers can be divided as they are starting into growth, using a sharp knife and making sure each section has a cluster of leaf buds. Dust the cut surfaces with sulphur powder before potting up.

Cyperus (umbrella plant) Propagated by seed or division in moderate heat.

Cytisus (florists' genista) Usually increased by cuttings of sideshoots with a slight heel taken in spring and inserted in pots, which are placed in a close warm frame. Seed may also be used and is sown in spring.

Dieffenbachia Increased by basal suckers or, like cordylines, by short stem pieces in brisk heat.

Dracaena (dragon plant) Propagated by stem pieces as for cordylines.

Erica (heath) Tip cuttings 2 cm (1 in) long are inserted in pots in sandy peat in spring. Keep close and use gentle heat.

Euphorbia (poinsettia) Cuttings are taken in early spring and struck in a close case with brisk heat.

Fatsia *F. japonica* can be raised from stem cuttings about 5 cm (2 in) long in spring. Insert in a close frame. Seeds sown singly in pots germinate at 18° C (65° F).

Ferns Division when re-potting, usually in spring, is generally adopted. The well-known *Asplenium bulbiferum* is increased by minute plantlets which develop on mature fronds. The fronds may be pegged down on light compost like a leaf cutting, or the plants may be taken off singly and carefully inserted round the sides of a pot.

Ficus (rubber plant) Propagate by soft, semi-ripe or leaf bud cuttings

in the greenhouse between spring and mid-summer. Large plants can also be air layered in summer.

Freesia Separate the offsets when re-potting in the autumn. Usually 6 to 8 corms are inserted in a 13 cm (5 in) pot. Easily raised from seed sown when ripe or in spring. Seedlings usually flower before they are one year old.

Fuchsia Usually raised from cuttings of soft young shoots. These are taken in spring or in autumn and inserted in pots of sandy compost. Seed may be sown in spring in pans and germinated in a temperature of about 15° C (60° F).

Gardenia These beautiful evergreen shrubs are increased by softwood or semi-ripe cuttings taken between spring and late summer in the greenhouse. Use sandy compost and keep the cuttings in close conditions.

Gesneria Can be increased naturally by division of the tubers. Leaf cuttings will strike when taken in summer. Seed is another method but requires care as it is extremely fine. Treat like begonias.

Gloxinia (sinningia) The gloxinia may be increased by seed which is

Left: Cabbages are raised in a seedbed and transplanted when large enough.

Above: All transplants must be watered in very thoroughly.

sown in pans of light compost in early spring. Basal cuttings are easily rooted in pots of sandy compost placed in a warm frame. Leaf cuttings provide another method.

Grevillea G. *robusta* is increased from seed sown in late winter in sandy compost at a temperature of about 21° C (70 ° F). Also propagated by semi-ripe cuttings in summer.

Heliotrope Cuttings of soft young shoots are inserted in spring in well-drained pots of light compost. Place in a close frame until rooted and then pot singly. Pinch the young plants two or three times to induce a bushy habit. Cuttings to be trained as standards are not stopped until they reach the required height. Also raised from seed.

Hibiscus (rose mallow) Root softwood cuttings in sandy compost in spring and early summer.

Hippeastrum Secure offsets when re-potting old bulbs in early spring. Seed should be sown in heat in early spring and potted on when necessary. Seedlings do not reach flowering age until they are three years old.

Hoya carnosa (wax flower) Semi-ripe cuttings are inserted into sandy compost in summer. Young shoots may also be layered during the summer by pegging them into pots of peat and sand.

Humea Sow seeds in pots or pans in mid-summer and place in a cold frame or glasshouse. Pot when large enough to handle.

Hydrangea The common hydrangea is increased by cuttings. Young shoots are taken in spring and inserted singly in small pots of sandy compost. Place in a warm frame until rooted and afterwards, grow in a cool glasshouse or frame. Pot as required.

Impatiens (balsam) Glasshouse species are easily raised from seed but special strains or varieties are readily propagated from softwood cuttings.

Ipomoea (morning glory) Annual species are raised from seed sown in early spring in a warm house. Perennial species are increased from cuttings made from sideshoots during the summer. Strike in a close case. *I. batatas* (the sweet potato) is increased by division of the tubers in late winter or from cuttings.

Jasminum (jasmine) Glasshouse species are propagated from soft young cuttings in spring; or by serpentine layering.

Kalanchoe Seed is raised in pots or pans of light compost in a temperature of 15–18° C (60–65° F). The seedlings need care in handling. Softwood cuttings and leaf cuttings strike readily in sandy soil.

Lachenalia (Cape cowslip) Bulbs increase naturally and are usually re-potted in late summer and planted 2–5 cm (1–2 in) apart. Seeds may also be sown in the spring in heat and are not difficult to germinate.

Lantana Propagated by seed sown in spring in a temperature of 21–24° C(70–75° F). Cuttings of young sideshoots inserted round the sides of a pot in spring strike in a warm case. Half-ripened cuttings taken at the end of summer and treated similarly will also root.

Lapageria Can be raised from seed in pans and kept in a warm house. Young shoots can also be layered indoors in spring or summer.

Monstera Like ficus, this plant can be increased by softwood cuttings, leaf bud cuttings or air layering, either in the home or greenhouse.

Nerine (Guernsey lily) Offsets are secured when re-potting, usually in late summer or early autumn and are potted in a 8–15 cm (3–6 in) pot.

Nerium (oleander) Cuttings made from semi-ripe wood are potted singly in sandy compost in spring or summer. Keep in a close frame until rooted.

Orchids These comprise a large and varied group of plants, and propagation must be related to the usual cultural practice for each kind. Many orchids can be increased by division, this being done when re-potting, which is normally undertaken just when growth is commencing in the spring. Genera treated in this way include Cattleyas, Calanthes, Cymbidiums, Cypripediums, Dendrobiums and Masdevallias. Propagation by seed is very difficult and undertaken by specialists only.

Palms These plants are usually increased by seed which is sown in spring or when available. Sow in well-drained pans or pots, which should be plunged in peat and kept in a propagating frame with a temperature of 21–24° C (70–75° F). A few species can be propagated from suckers, which are potted in spring.

Passiflora (passion flower) Cuttings 10–15 cm (4–6 in) long with a heel are made from young shoots in spring. These are inserted in a warm propagating frame. Serpentine layering and seed are other methods.

Pelargonium (geranium) The different types are raised from cuttings which are cut off below a joint and are made 5–10 cm (2–4 in) long. Pot singly in small pots using sandy compost and place on a glasshouse staging. When rooted, the young plants are potted on as required. Some species and varieties may be raised from seed.

Peperomia Cuttings of short stem pieces with a single joint will root in pots stood on a glasshouse staging. Under closer more humid conditions damping-off is probable. Seed is another method.

Philodendron Increase as for *Monstera*.

Plumbago (Cape leadwort) Easily propagated by soft or semi-ripe cuttings with bottom heat.

Primula Seed is the normal method of increasing glasshouse primulas and is usually sown in spring for winter flowering. *P. malacoides*, however, is sometimes sown in mid-summer to provide flowers in spring. Sow the seeds in pots and prick off the seedlings when they are large enough to handle. Later, they are potted into 8 cm (3 in) pots and finally into the 13 cm (5 in) size. A temperature of 13–15° C (55–60° F) is suitable for seed germination and 10–13° C (50–55°F) for growing the seedlings. Double-flowered primulas are raised from cuttings or division.

Saintpaulia (African violet) Usually raised from leaf cuttings taken throughout spring and summer. Can also be raised from seed sown in early spring and kept in a temperature of 18–21° C (65–70° F). Keep young plants well shaded.

Salvia *S. splendens* is usually raised from seed which is sown in the spring in pots or boxes. Germinate in a temperature of 15–18° C (60–65° F) and pot singly in 8 cm (3 in) pots when the seedlings are large enough to handle. Most salvias may be increased from cuttings. These are secured from plants of the previous year retained over winter in mild heat and started into growth in spring. Insert the cuttings in pots which are placed in a propagating frame.

Sansevieria Easily increased by suckers. Also by leaf cuttings, the long leaves being cut into lengths of about 8 cm (3 in) which are inserted in sandy compost in heat.

Saxifraga sarmentosa (mother of thousands) This species increases naturally by creeping stems or runners. From these, the plantlets are removed and inserted in small pots.

Schizanthus (butterfly flower) For summer flowering, sow the seeds in early spring in a warm house. Late summer is the time to sow when the plants are required to bloom in the spring. Prick the seedlings off and pot as necessary.

Selaginella (creeping moss) Cuttings inserted in pots strike readily in a close frame. When rooted, several cuttings may be potted together to form a single specimen. Division when re-potting is another method of increase.

Solanum *S. capsicastrum*, the popular winter cherry, is usually increased from seed which is sown in late winter in heat. Prick off the seedlings and pot when they are large enough, first into 8 cm (3 in) pots and finally into the 13 cm (5 in) size. Pinch the plants to induce a bushy habit. The winter cherry may also be increased by cuttings inserted in pots placed in a propagating frame in early spring. The climbing species such as *S. jasminoides* are increased from soft young cuttings in spring struck in pots in a close case.

Strelitzia (bird of paradise flower) May be raised from seed sown from mid-winter through the spring kept in a high temperature, preferably with bottom heat. Old plants may be divided when being re-potted in spring.

Streptocarpus (Cape primrose) May be raised from seed sown in winter for flowering the following winter or in summer for summer flowering. Germinate in a temperature of 15–18° C (60–65° F). Leaf cuttings, using sections of healthy leaves, will grow if inserted in sandy compost in spring and summer. Established plants can be divided into separate crowns after flowering.

Streptosolen *S. jamesonii* is increased by cuttings made from soft young shoots in spring or summer and inserted in a close case. Rooted cuttings are potted singly and pinched once or twice to encourage a bushy habit.

Thunbergia Propagate by seed sown in heat or by softwood cuttings in a warm frame.

Torenia Usually raised from seed sown in spring and treated as a glasshouse annual. It may, however, be increased by cuttings in a close case.

Tradescantia (spiderwort) Easily increased by cuttings of young shoots inserted in a close frame.

Zantedeschia (arum lily) Re-potting and division are usually done at the end of summer. Suckers which are readily produced are potted singly and started in mild heat. Seed provides another method and may be sown in spring in a warm glasshouse.

Zebrina Easily increased from cuttings in moderately warm conditions at any time.

INDEX

Individual plants are listed alphabetically between pp. 48–95.
See also entries for Annuals and Biennials.